RADIOGRAPHY IN THE EARTH SCIENCES AND SOIL MECHANICS

Monographs in Geoscience
General Editor: Rhodes W. Fairbridge
Department of Geology, Columbia University, New York City

RADIOGRAPHY IN THE EARTH SCIENCES AND SOIL MECHANICS

E. L. KRINITZSKY

Waterways Experiment Station
Corps of Engineers, Vicksburg
Mississippi

℗ SPRINGER SCIENCE+BUSINESS MEDIA, LLC 1970

Library of Congress Catalog Card Number 70-107539
SBN 306-30448-1

ISBN 978-1-4684-1805-7   ISBN 978-1-4684-1803-3 (eBook)
DOI 10.1007/978-1-4684-1803-3

© 1970 Springer Science+Business Media New York
Originally published by Plenum Press New York in 1970
Softcover reprint of the hardcover 1st edition 1970

ACKNOWLEDGMENTS

The preparation of this book was made possible by the interest and assistance of a great many people. I am particularly grateful to James M. Coleman and Sherwood M. Gagliano of the Coastal Studies Institute, Louisiana State University, for much valuable information. Original materials were contributed by William K. Hamblin, Adrian F. Richards, Arnold F. Bouma, James M. Coleman, James J. O'Dea, Daniel J. Stanley, William M. Kirkpatrick, James A. Wolleben, George Gornak, Eli I. Robinsky, and others. Charles R. Kolb, Stanley J. Johnson, Edward B. Perry, and A. A. Maxwell assisted with reading of the manuscript.

I am also deeply grateful to the U. S. Army Engineer Waterways Experiment Station (WES), where I was employed. Many of the illustrations were taken from studies conducted at WES in the field of radiological research. These and other materials are in various stages of being published in WES reports. Permission to use the results of these studies was granted by the Director, WES.

Following are some of the persons who contributed to WES projects in radiography: tests in the radiological laboratory were done by David P. Ripley, Robert O. Pichulo, F. L. Smith, Anna Leese Marshall, and Jack T. Lewis. Much of the photographic work was done by Francis B. Gautier, Ryland B. Rudd, Joseph H. Lindsey, and others. I would like to mention that the high standards which they apply to their photographic work made it possible, earlier, for us to exhibit a group of our radiographs at the Museum of Modern Art in New York City. Bibliographic assistance was contributed by Marie Spivey, Estelle S. Sigler, Elizabeth Garrett, and others. Editing and preparation of reports was done by Mary B. Pikul and others. To all these people, and to many who have gone unnamed, I wish to express my deep appreciation.

CONTENTS

Chapter 1

INTRODUCTION

Radiography, the use of penetrating radiation to produce shadow images of the internal structure of materials, has been with us since Roentgen made his discovery of x rays in 1895. However, applications of radiography in the earth sciences and in the related field of soils engineering have, until recently, been slow to develop.

Brühl reported optimistically on applications in paleontology as early as 1896 and there have been additional reports through the years. However, very few paleontologists adopted the method and the significant literature is relatively restricted. In soil mechanics, Gerber observed the movement of lead pellets in sand during a plate-bearing test as early as 1929. Gradually, radiography was applied to other tests including those on footings, compaction of soils, strain in sand, effects of pile penetration, and displacements under moving wheel loads. Recently, such work has broadened into much varied and sophisticated research.

Applications in geology may be dated to Hamblin's work on rocks reported in 1962. His demonstration that many fine textural and structural details can be observed in slices of rock led to experimentation by others on unconsolidated sediments and soils. Work is now expanding at an unprecedented rate. In some operations, such as the logging of oceanographic cores, it is already a routine process.

The advantages of radiography lie in its nondestructive nature and its ability to reveal features that sometimes cannot be seen in any other way. Its applications in the earth sciences and soil mechanics will be examined in the following sections of this book.

Successful radiographic work with earth materials has, for practical purposes, been done entirely with x rays. However, some minor applications involve gamma rays and there are possible future developments in the application of neutron radiation. Hence, emphasis here will be mostly on x rays, but with minor reference to other techniques; gamma ray and neutron sondes and metering devices are not within the scope of this volume.

For technical detail on the principles of radiation, and on the construction and use of radiation equipment, the reader is referred to the excellent source volumes listed at the end of Chapter 2. The present volume will treat only the most essential elements of radiation theory and of radiation equipment.

Understandably, this volume will be devoted to demonstrating successes in the applications of radiography. A word of caution is in order. There are earth materials for which radiography is not as successful as other techniques. Particularly, this occurs among dense but featureless rocks and dense rocks of considerable complexity but of uniform mineral composition. The latter condition is obtained frequently in crystalline limestones, for which acetate peels may show more than can be seen in radiographs. Even where detail is registered in a radiograph, there are instances where the detail may show up equally effectively on an ordinary cut or polished surface. K. H. Wolf (1967) mentioned that radiographs were unsuccessful in his work with dense uniform limestones. Also, Ali and Weiss (1968) demonstrated that transmitted infrared photography revealed more detail in a specific dense limestone than was observable in a radiograph. However, in their experiments they worked with samples 2 mm thick; these specimens may have been too thin for optimum use of radiography, though the thickness was proper for infrared photography.

It must be kept in mind that there are limits to the applicability of radiography and that radiography should be used in combination with other techniques of the geological and soils engineering laboratories.

BIBLIOGRAPHY

Ali, S. A. and M. P. Weiss (1968). Transmitted infrared photography: Cincinnatian lime-
     stones, *J. Sed. Petr.* **38** (4): 1350–1354.
Wolf, K. H., A. J. Easton, and S. Warne (1967). Techniques of examining and analyzing
     carbonate skeletons, minerals and rocks, *in*: George V. Chilingar, Harold J. Bissell, and
     Rhodes W. Fairbridge (eds), *Carbonate Rocks*, Elsevier, New York, Chapter 8, p. 289.

Chapter 2

ELEMENTS OF RADIOGRAPHY

Persons who are newly concerned with radiography may obtain an excellent introduction to the subject by reviewing some or all of the books in the general reading list at the end of this chapter. The following sections outline only the essential elements of radiographic theory.

CHARACTERISTICS OF X RADIATION AND GAMMA RADIATION

X rays are generated when a stream of high-energy electrons are slowed upon striking a target. Alterations in the orbital shells of electrons on the

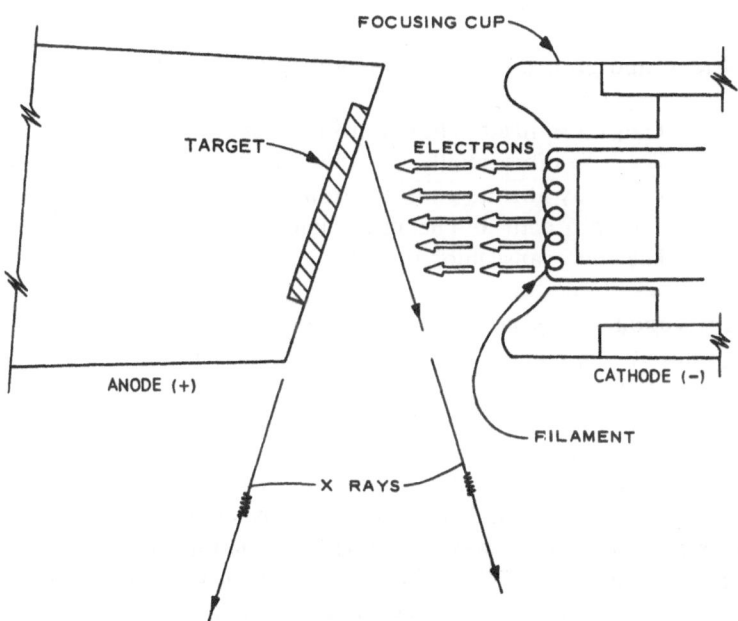

Fig. 2-1. Schematic view of x-ray emission.

3

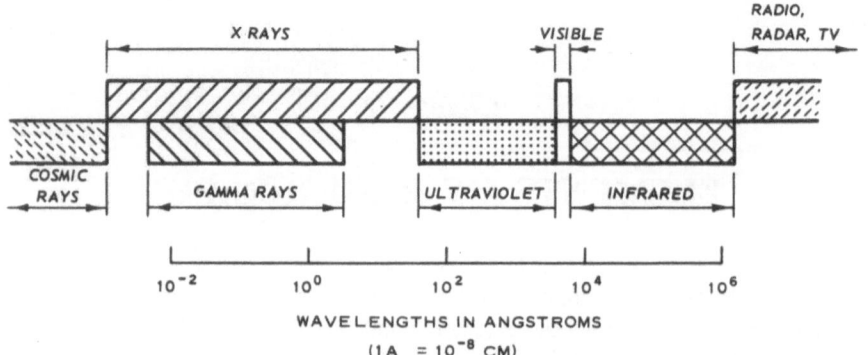

Fig. 2-2. Position of x rays and gamma rays in the electromagnetic spectrum.

target give rise to photons or quanta of energy that are x rays. The production of x rays is shown schematically in Fig. 2-1. The x rays produced are in the portion of the spectrum indicated in Fig. 2-2.

Gamma rays have the same wavelengths as x rays, the only difference being in their mode of origin. Gamma rays are produced as a part of the process whereby radioactive isotopes break up their nuclei and give off radiation.

X rays, and gamma rays, have the following properties:

1. They travel in straight lines with the speed of light
2. They are differentially absorbed by materials they pass through
3. They may liberate photoelectrons
4. They can activate chemical changes in photographic emulsions
5. They may cause fluorescence

Absorption is dependent upon the wavelength of radiation and the density and atomic number of the absorbing material.

Wavelength is dependent on the voltage (generally expressed as kilovoltage, kV) applied to the x-ray tube. As the kV is raised, x rays of shorter wavelength and more penetrating power are added. Inversely, a lowering of the kV produces proportionately longer wavelengths and less penetrating power. Long-wavelength radiation requires greater time and produces increased scatter; the shorter wavelengths (hard radiation) are therefore desirable for thick specimens. With thin specimens long wavelengths (soft radiation) are superior in achieving delicate registration of detail.

Density changes in a soil or rock may result from alternations in the packing or structure of mineral grains. Variations in atomic number may result from changes in the quantity and composition of cementing material from place to place in a specimen. These changes are registered by differential absorption of x rays. In many cases the features that are revealed may not be recognizable by other means.

SCATTERED RADIATION

Some x rays emerge from an absorber in directions different from those at which they entered. This is scattered radiation and scattering of the same beam may be a multiple occurrence.

The complex nature of scatter is suggested in the schematic listing in Fig. 2-3. A photoelectron is produced when a photon is absorbed by an atom and one of the bound electrons in the atom is ejected. A photon may be deflected and travel in an altered direction with decreased energy, the electron causing the deflection recoiling and being ejected in a different direction. This is called the Compton effect. A high-energy photon may be absorbed in an atom and a pair of electrons, one charged positively and one negatively, may be ejected. Figure 2-3 also itemizes absorption and passage of primary radiation, fluorescence, and other effects.

Scatter, when it occurs excessively, tends to fog x-ray film and possibly obscure details that otherwise would be registered. Unfortunately, scatter can never be entirely eliminated but it can be mitigated in a number of ways. Masks and shields of lead may be used to concentrate primary radiation and shield out scatter and, since scatter arises also from the specimen itself, it is good practice at times to restrict the x-ray beam to a limited area of a specimen. Under some circumstances a filter placed near the x-ray tube is helpful. The filter is a metal sheet that stops soft radiation but permits hard radiation to pass. This stopped radiation would normally not penetrate a large object but would penetrate peripheral parts excessively and induce undercutting.

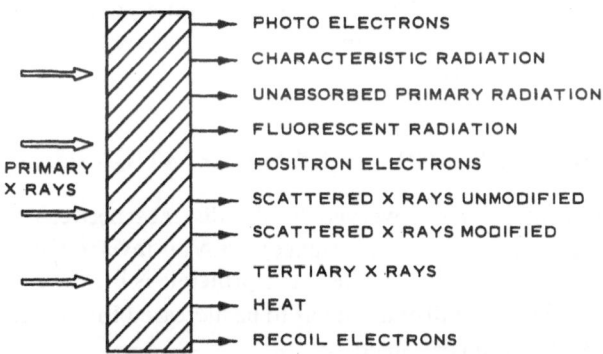

Fig. 2-3. Modifications of primary x rays.

By removal excessive subject contrast is reduced and the image is improved in its peripheral parts.

A special grid of very thin lead lines placed against the film on the radiation side may reduce fogging. These lead lines are so fine that they impart a halftone effect to the film and are hardly noticeable yet they absorb scattered radiation and allow direct rays to be registered. The grids are specially designed for given focal distances and should be used according to their specifications. Backing the film container with a lead sheet helps to eliminate backscatter that may come from below the film.

Problems of scatter are usually associated with thick specimens. Where observations can be made on relatively thin slices of soil or rock, energy levels of radiation can be kept low enough so that scatter is only a minor problem or no problem at all. For this reason most of the illustrations in this volume were made from soil and rock slices $3/8$ in. thick.

ABSORPTION

Radiation scatter involves changes in photon direction. True absorption occurs through the conversion of photons into increased kinetic energy of electrons within matter that is radiated. To a much lesser extent, radiation photons are also absorbed by interaction with the nuclei of atoms.

Mass-absorption coefficients are available individually for the major elements in terms of varying energy levels of radiation. These coefficients are based on measurements of scattering, photoelectric effects, pair formation, and mass absorption. Absorption coefficients for compounds such as quartz, kaolinite, or other minerals that make up earth materials can be calculated by summing the coefficients of constituent elements. Such calculations have applicability where samples are prepared artificially, as for soils tests, under controlled conditions and are composed of known constituents. Applications will be discussed further in Chapter 11; however, for earth materials in general there is such heterogeneity in composition and structure that the precision of these calculations is rarely applicable.

DISTANCE AND INTENSITY RELATIONSHIPS

Radiation intensity is governed by the distance between the tube, or source, and the specimen (focal distance), with an inverse variation that is based on the square of this distance. The principle is illustrated in Fig. 2-4. The radiation effective at distance d has to be increased four times in order to be equally effective if d is doubled.

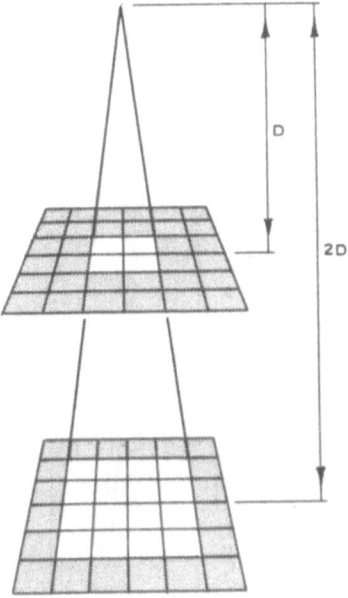

Fig. 2-4. The inverse-square law.

The inverse-square law is expressed algebraically as follows:

$$\frac{I_1}{I_2} = \frac{d_2^2}{d_1^2}$$

where I_1 and I_2 are the radiation intensities at focal distances d_1 and d_2, respectively.

The amount of radiation intensity I in a given exposure is measured as a combination of amperage (generally expressed in milliamperes, mA) and time. The amperage for a given exposure may be varied in an inverse proportion to the time. Thus,

$$\frac{M_1}{M_2} = \frac{t_2}{t_1}$$

where M is amperage and t is exposure time.

The relation of amperage to focal distance is

$$\frac{M_1}{M_2} = \frac{d_1^2}{d_2^2}$$

The relation of time to focal distance is

$$\frac{t_1}{t_2} = \frac{d_1^2}{d_2^2}$$

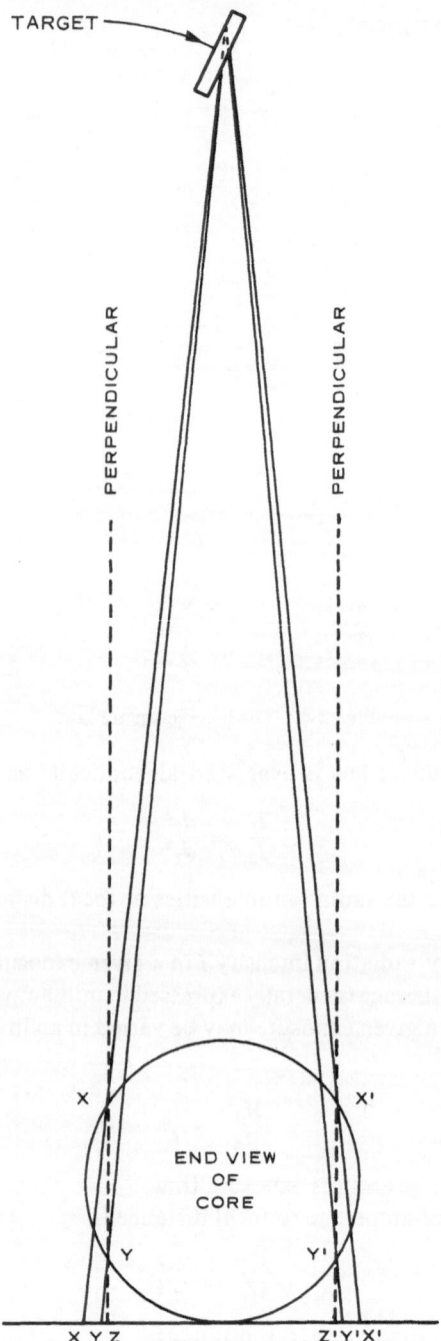

Fig. 2-5. Distortion of x-ray image by conical spreading of rays. Note displacements of points x and y in relation to projection z. Distortion is greatest on the periphery and negligible in the center.

The three equations may be combined to form an exposure factor E_f:

$$E_f = \frac{Mt}{d^2}$$

When E_f and the tube voltage are kept at predetermined levels radiographs of equal quality may be obtained. Adjustments are made in amperage, time, and focal distance; a further adjustment in the time of exposure may be made to obtain a desired film density. Suitable corrections are indicated in Chapter 3.

Application of these relationships is essential in the use of radiographic exposure charts discussed in the following chapter.

GEOMETRIC FACTORS

Radiation is in the form of a cone emanating from its source. Thus, the image that is registered is subject to distortion depending on the shape of the sample and the degree to which the rays diverge. Figure 2-5 shows the distortion that takes place in a typical radiograph of a soil core. Instead of registering points x and y at z, x and y are at different positions with relation to z and with relation to each other. Such distortion is greatest in the periphery but it is negligible in the center of the specimen.

The diagram presupposes that the conical radiation comes from a point source. Actually the source on the target in the x-ray tube is an area called the focal spot, the size varying with the design and power of the tube. A small focal spot permits sharper registration or definition of radiation. The more powerful tubes have larger focal spots; consequently they produce a broadening of the shadows and a corresponding loss of definition.

In practice it is preferable to have the greatest feasible distance between the x-ray source and the object with the film as close to the object as possible to prevent further spreading. The shape of the object presents fewest problems when it is a thin, flat slice of soil or rock.

In certain model studies, for example, when the movement of lead pellets

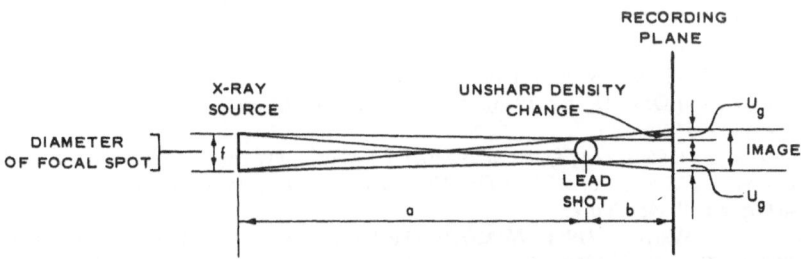

Fig. 2-6. Effect of focal spot and focal distances on geometric unsharpness of film image.

is being traced in a soil under deformation, it is desirable to know the degree of unsharpness resulting from the broadening of shadows. Clauser (1952), Halmshaw (1966), and Arthur and Shamash (1967) present the following relationship (see Fig. 2-6):

$$U_g = f\left(\frac{b}{a}\right)$$

U_g is geometric unsharpness, f is the diameter of the focal spot, and a and b are the distances indicated in Fig. 2-6.

NEUTRON RADIATION

The absorption of neutrons by matter is comparable to that of x rays and gamma rays at high energy levels. However, neutrons are either absorbed by a nucleus to form a different nucleus or they are scattered by the nucleus, and relative absorption by the various elements is entirely different from that for x and gamma rays.

Of particular interest in the earth sciences is the high relative absorption of neutrons by hydrogen, neutrons thereby being sensitive indicators of water and hydrocarbons. Thus far their application has been limited to sondes used in boreholes for soil investigations or in wells drilled for petroleum exploration, and in surface meters designed to determine soil moisture. Pictographs of moisture penetrations in rock and soil could aid research in a great many facets of the mechanics of these materials. Such work has not been done to date, hindered by the need for large installations in order to generate suitable neutron radiation.

BIBLIOGRAPHY

Arthur, J.R.F. and S. J. Shamash (1967). A note on the accuracy of displacement measurements in soils, *Civil Eng. and Public Works Review* **62** (729): 455–456.

Clark, George L. (1963). *The Encyclopedia of X-Rays and Gamma Rays*, Reinhold, New York.

Clauser, H. R. (1952). *Practical Radiography for Industry*, Reinhold, New York, pp. 100–103.

Eastman Kodak Company (1957). *Radiography in Modern Industry*, second edition w/supplements, Rochester, N. Y., 136 pp.

Guinier, A. and D. L. Dexter (1963). *X-Ray Studies of Materials*, Interscience Publishers, New York, 156 pp.

Halmshaw, R. (1966). *The Physics of Industrial Radiography*, Heywood, London. Kaelble, E. F. (1967). *Handbook of X-Rays for Diffraction, Emission, Absorption, and Microscopy*, McGraw-Hill, New York.

McGonnagle, Warren J. (1961). *Nondestructive Testing*, McGraw-Hill, New York, 445 pp.

McMaster, Robert C (1959). *Nondestructive Testing Handbook*, two vols., The Ronald Press, New York.

Price, B. T., C. C. Horton, and K. T. Spinney (1957). *Radiation Shielding*, Pergamon Press, London, 350 pp.

Quality and Reliability Assurance Handbook H55 (1966). *Nondestructive Testing Series, Radiography*, Office of the Asst. Secty. of Defense (Installations and Logistics), Washington, D. C., 202 pp.

Richardson, Harry D., A. L. Bertrand, and D. E. Shipp (1964). *Industrial Radiography Manual*, prepared for U. S. Atomic Energy Comm., Div. of Nuclear Education and Training, contract AT-(40-1)-3112, Washington, D. C.

Trillat, Jean-Jacques (1959). *Exploring the Structure of Matter*, transl. by F. W. Kent, George Allen & Unwin Ltd., London.

Chapter 3

RADIOGRAPHIC IMAGES AND SYMBOLS

In radiography picture forming involves either fluoroscopy or registration on special photographic film. Images can also be recorded on photographic paper, Polaroid film and paper, and on various color films; however, such applications are exceptional.

FLUOROSCOPY

Fluoroscopy makes use of the ability of certain materials to fluoresce when they absorb x rays or gamma rays. Sensitivity and resolving power is not as great as with radiographic film and higher tube voltages must be used. Because the fluorescent screen does not accumulate the effects of radiation the way film does, fluoroscopy decreases in resolving power as the material under examination thickens. With soil or rock, thicknesses up to 3 in. can be viewed successfully, 5-in. thicknesses with much less success, and greater thicknesses with a probable insufficient clearness.

Most fluoroscopic screens are made with zinc–cadmium preparations and are used with conventional, low-voltage x-ray sources, and photographs can be made of them. Variables affecting Screen performance include material thickness, chemicals, crystal sizes, and x-ray intensity at a given voltage. At voltages above 160 kV fluoroscopy decreases in effectiveness and even under optimum conditions there is an unsharpness that prevents attaining a knife-edged accuracy in viewing details. The operator or viewer must accustom his eyesight to the light from the viewing frame for 15 or 20 minutes before he can make observations with maximum ability, and must also avoid tiring his eyesight; consequently, he is limited to short periods of viewing.

Despite these handicaps fluoroscopy can be highly useful in earth studies. It permits quick scanning and culling of samples and a rapid check on soil specimens and models before they are subjected to testing.

RADIOGRAPHIC FILM

Radiographic film is composed of an emulsion—gelatin containing a silver salt—that is spread on a cellulose derivative base. The emulsion is usually coated on both sides of the base in thicknesses of about 0.001 in., although special-purpose films have emulsion on one side of the base only. The emulsion and base is sandwiched between covering material to form the film unit. Radiographic film has neither front nor back, except as determined by the lead numerals or letters used to identify samples, and can be exposed, viewed, and printed from either side.

Various types of film are produced that incorporate differences in speed and differences in registration of detail. As with regular photography, high speeds provide less detail than low speeds. Table 3-1 compares film types currently in use.

Film may be loaded into cassettes in a darkroom or it may be purchased prepackaged in lightproof envelopes. After exposure film may be accumulated and processed in groups either manually or by automatic machines, using chemicals specified by the manufacturers.

Film density (film darkening) is a function of the exposure time and the energy of radiation and ranges from 0 to 5. Figure 3-1 shows the variation in film density that results from increases in exposure for a group of Kodak industrial x-ray films. In general the changes follow logarithmic functions and the curves can be used to estimate changes in exposure time needed to achieve desired film densities. For practical purposes the characteristic curves are independent of the specific voltage used and the x-ray wavelength distribution.

The time used in developing film also affects its density, with longer time producing greater density. It is desirable to standardize this part of the procedure and to check the development process from time to time

Table 3-1. Comparability Chart for Radiographic Film
(After Fraser and James, 1969)

Kodak	Ferrania	Gevaert	Ansco	Du Pont	Rel. Speed	Rel. Detail
—	—	D2*	—	—	1/2	Highest
R	—	D2	HD	—	1	Very high
M	1-Gamma	D4	B	510	4	High
AA	IC2	D7	A	506	16	Good
KK	ID	D10	C	—	64	Medium
F	IS2	S	D	504	Very high	Fair
—	—	Super S	—	—	Highest	Fair

* Single Emulsion. Kodak Type T film has a relative speed of 8.

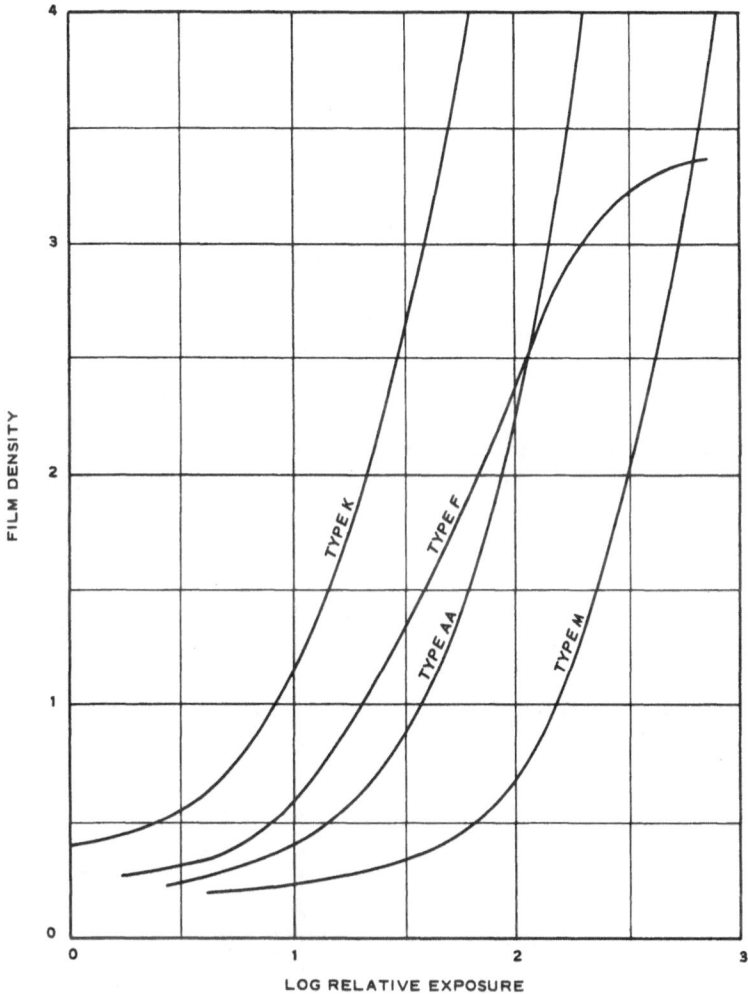

Fig. 3-1. Characteristic exposure curves for Kodak industrial x-ray films, types AA, F, K, and M (after Kodak, 1957).

with prepared film strips. Film densities of 0.75 to 1.5 are easiest to scan and may be looked at with almost any convenient light source including overhead fixtures. However, denser film usually holds greater detail. Densities up to about 4.5 can be observed effectively with the aid of high-intensity back-lighting; several illumination sources are available for this purpose.

A tiny portion of x-ray energy, less than 1% is absorbed by the film. Thus, several sheets of film, placed one on top of the other, can be exposed at once if multiple copies of the image are desired. They will all be of equal quality.

If a soil specimen is composed of layers that vary greatly in density, such as a peat layer in a sand, it is inevitable that some parts of the film will become excessively darkened while other parts remain light. Two shots may be needed, using different exposures and providing the picture on two pieces of film. Sometimes one exposure with two sheets of film of different speeds can accomplish the same purpose.

The image quality on film can, in certain cases, be enhanced by the use of lead and fluorescent intensifying screens.

Lead Screens.   Since less than 1 % of radiation affects film it is practicable to use part of the radiation to excite other emission that will register on film. Lead foil between 0.004 and 0.006 in. thick can be placed on top of film, with thicker foil behind the film to cut backscatter. Radiation will cause a release of photoelectrons from the lead and their differential emission is registered on the film. The lead also absorbs long-wavelength scatter.

Lead foil screens may be used where the voltage is greater than 130 kV and the materials are thick and dense, as with rock cores that are 4 in. or greater in diameter.

Fluorescent Screens.   Certain chemicals, such as calcium tungstate and barium lead sulfate, have the ability to absorb x rays and gamma rays

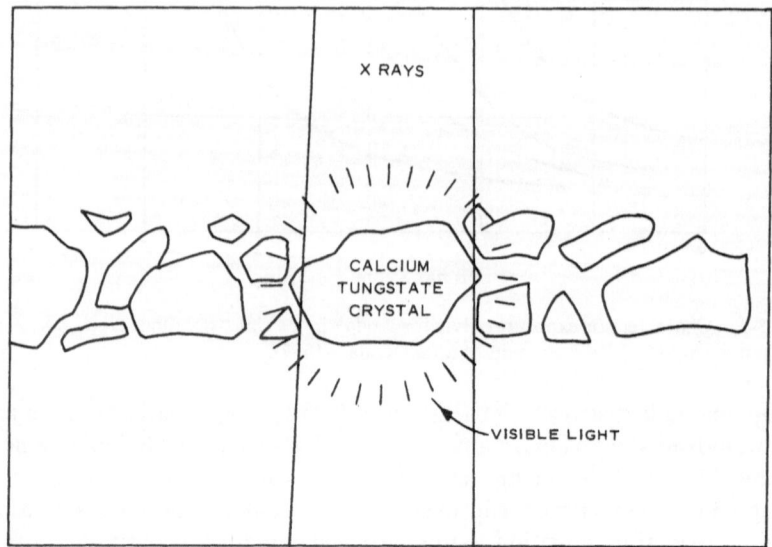

Fig. 3-2.   Schematic view showing light spread caused by glow of calcium tungstate crystal in a fluorescent screen (after Kodak, 1957).

and immediately emit light. Screens can be made utilizing these chemicals by binding them in a fine-powder form onto cardboard, and the film is clamped between a pair of these screens for exposure. In medical radiography the exposure is 1/15 to 1/30 with screens as without them, the intensification thus being 15 to 30 times. This technique can be useful with dense rock and with thicknesses of 5 in. or greater, and radiographs at 250 kV can be made with reasonable exposure times. However, because of light spreading there is a loss of sharpness of resolution, as shown in Fig. 3-2, and fluorescent screens do not provide the same quality or definition obtained in straight radiographs or radiographs using lead screens.

Once a radiograph is made on a piece of film, the resulting picture can be studied and interpretation can be made. Usually there is no need to reproduce the radiograph, but there are cases where representative illustrations are needed, as in reports and in this volume. It is good practice to make a film negative of the radiograph and make prints from the negative; the picture then produced has the same lights and darks found in the original radiograph and the same details can be described that would be seen by looking at the radiograph. This procedure has been followed in preparing the illustrations in this volume, unless stated otherwise.

Ordinary radiographic film is generally much coarser grained than photographic film and even fine detail cannot be enlarged more than five diameters without becoming grainy. When necessary, portions of radiographs can be cut and mounted between glass plates for use in lantern slide projectors. Larger portions of radiographs or even entire radiographs can be exhibited by use of an overhead projector (see Brixner and Richards, 1967).

EXPOSURE CHARTS

Exposure is determined by the variables input energy (as a combination of voltage (kV) and amperage (mA)), exposure time, specimen thickness, focal distance, and type of film or viewing surface. At times it may be necessary to achieve satisfactory working combinations through trial and error, which can be tedious and wasteful. Some data are available that may serve as guides for exposures but they must be used with some caution; the same measured energy input in various x-ray machines produces different x-ray intensities and differences in circuits produce variations in peak potentials during operation. Thus, the experience of other workers often must be modified by an unknown correction factor. Such a relation is evident in exposure guides produced recently by Hamblin (1967) and Fraser and James (1969).

Hamblin provided a group of exposures for eight rock types (Figs. 3-3) and 3-4), and Fraser and James published charts for three rocks and one

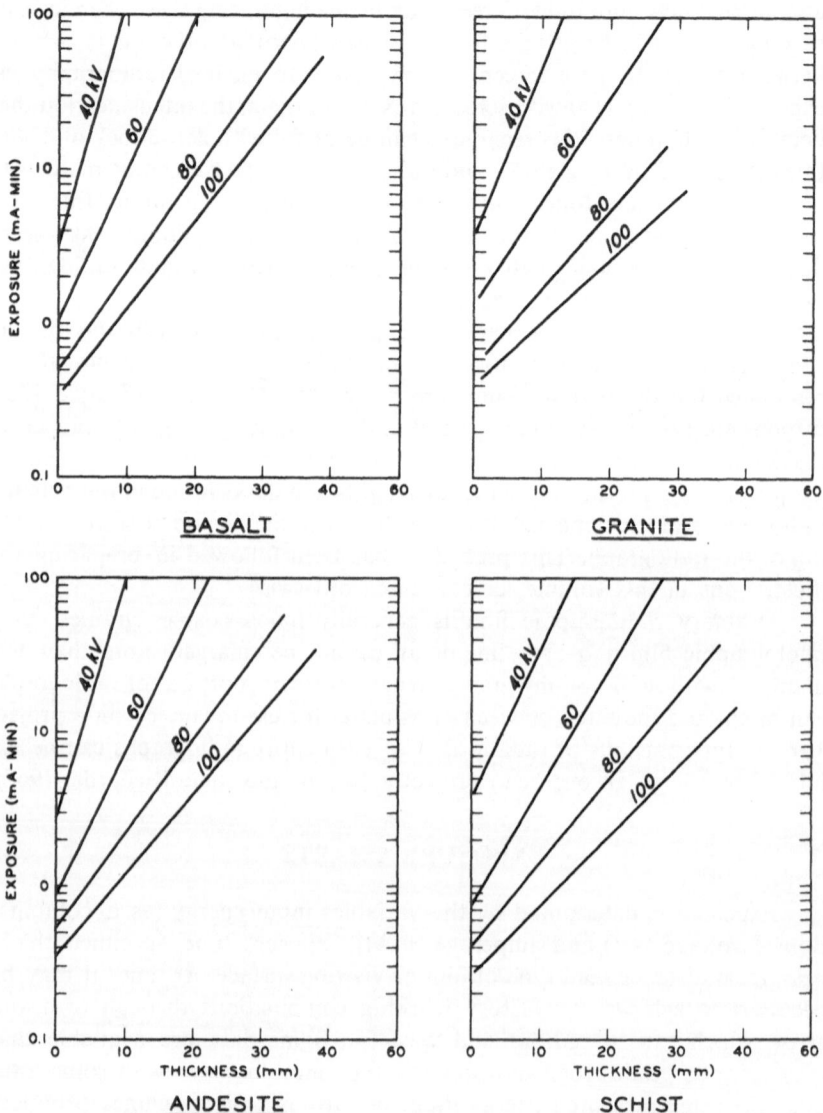

Fig. 3-3. Exposure charts for igneous and metamorphic rocks. Kodak type AA film; 40-in. focal distance (after Hamblin, 1967).

unconsolidated sediment (Fig. 3-5). It may be noted that where charts of lithologically similar material were provided (compare their limestones, sandstones, and shales) the exposure values differ considerably. The state of the art at this time is such that these differences are not resolvable by trans-

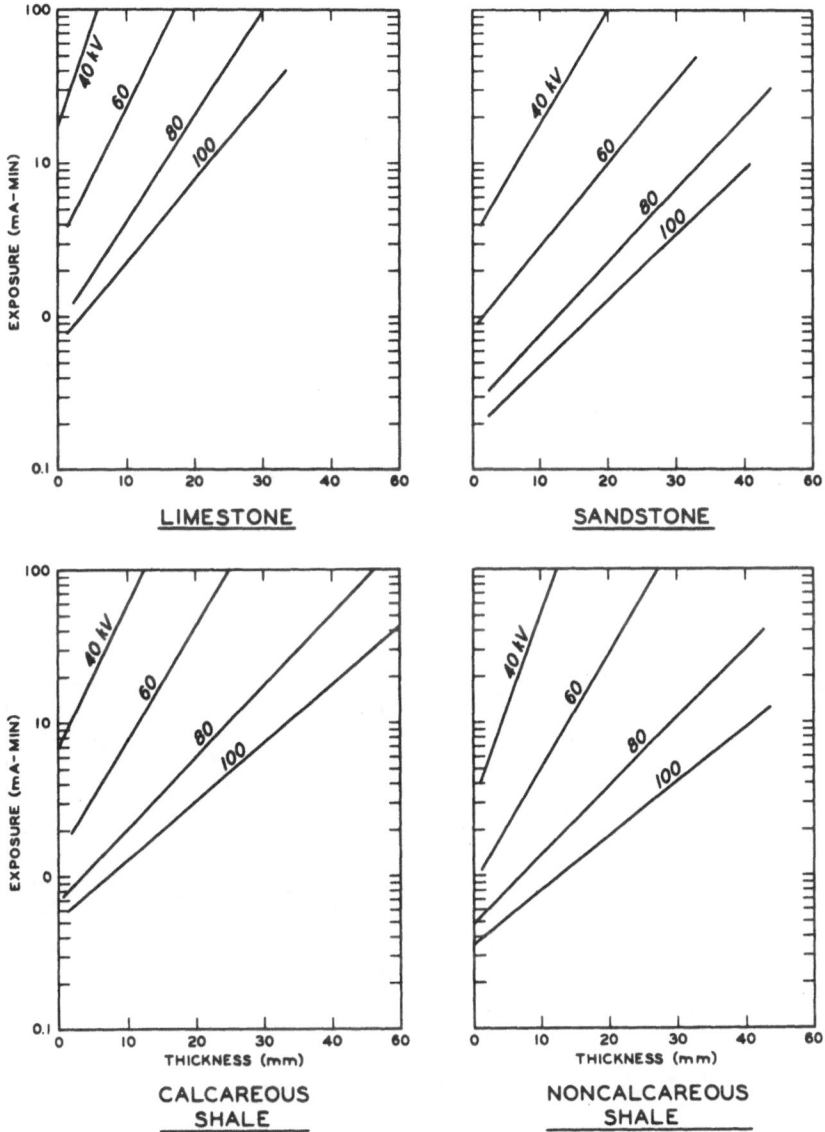

Fig. 3-4. Exposure charts for sedimentary rocks. Kodak type AA film; 40-in. focal distance (after Hamblin, 1967).

ference factors related to circuit designs. However, these charts are useful nonetheless for comparisons between materials and they may be used as starting points for testing, although each operator must work out his own exposure values.

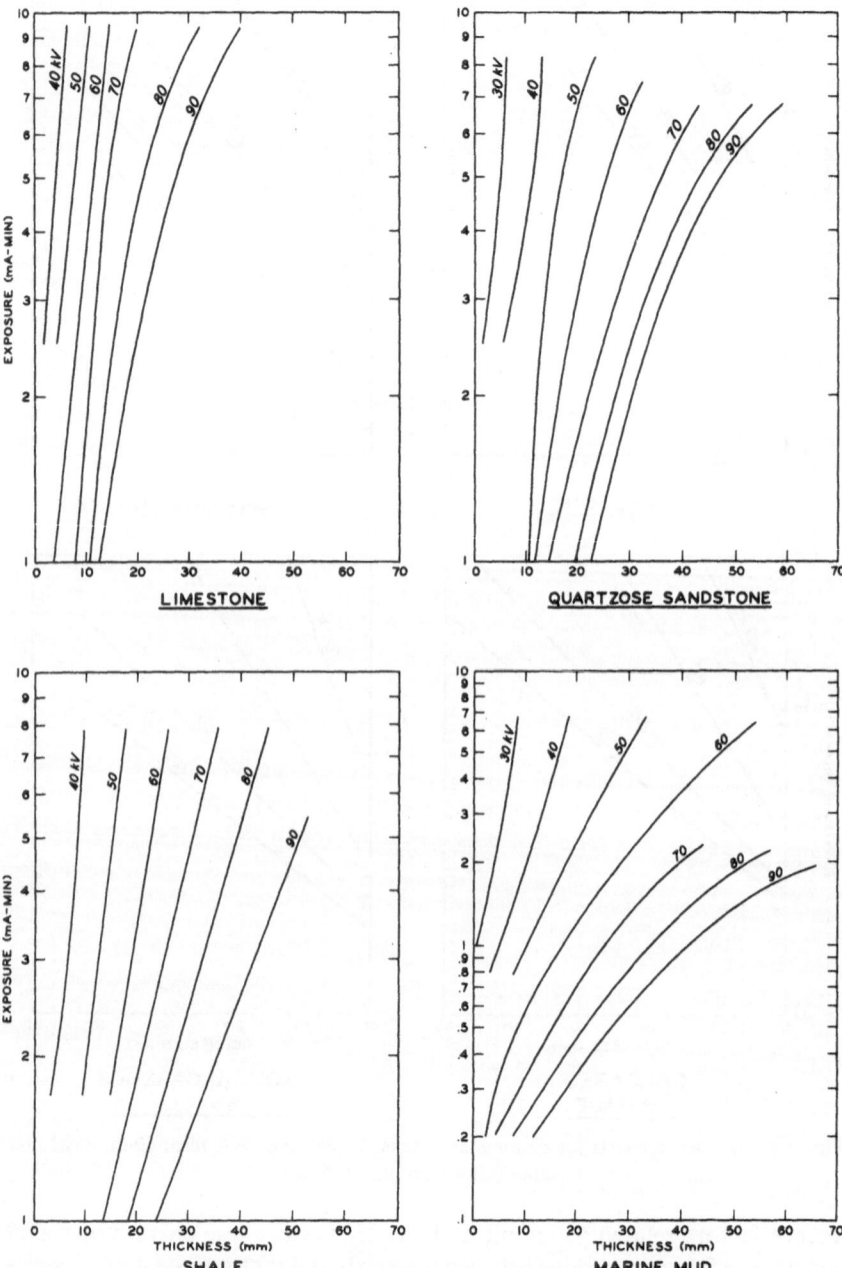

Fig. 3-5. Exposure charts for sedimentary materials. Kodak type AA film; 40-in. focal distance (after Fraser and James, 1969).

It may be noted that Hamblin's isopotential lines are straight while Fraser and James' are curved. Fraser and James suggest that dense materials, having no pore space, attenuate x-ray beams as a function of thickness and linear absorption coefficient, producing straight isopotential lines. Rock materials with pore space cause changes in the linear absorption coefficient as the beams pass from rock materials to interstitial space. Porosity also presents many surfaces which aid the dissipation of energy through scatter. These effects cause the curvature seen in the isopotential lines. Their reasoning appears to be confirmed by the straight lines measured in their shale sample (Fig. 3-5) as opposed to varying degrees of curvature in their other samples. Their shale is a dense fissile black shale from the Modesta formation of Pennsylvanian age in northwestern Illinois, their limestone is a lithographic limestone from the Davenport member of the Wapsipinicon formation of northwestern Illinois, and their sandstone is the St. Peter sandstone of Ordovician age from northern Illinois. The marine mud is unconsolidated mud from the floor of the Wilkinson Basin, Gulf of Maine. The most radical curvature is in the least consolidated material.

Fraser and James further suggest that the degree of curvature of the isopotential lines is a measure of the degree of lithification. Rocks of differing stages of lithification but of the same materials can be measured by proportional adjustments in the exposure guides.

In tests on iron D'Adler-Racz (1966) compared three types of x-ray circuits (the self rectified, constant potential, and rectified pulsating) to determine how they affected image quality. He concluded that there does not seem to be any significant difference in radiographic quality caused by type of equipment used.

PRINT ENHANCEMENT

Radiographs often have excessive contrast. Though they can be viewed and interpreted without great difficulty, they frequently need to be enhanced in order to be suitable for use as illustrations in reports. Photographic paper can be selected for improvement in image—soft paper for high-contrast film and high-contrast paper for low-contrast film—but most of the time this is not enough. Further enhancement of image is needed. Figures 3-6 to 3-8 show comparisons between possible means of printmaking from a high-contrast radiograph. The sample (see Fig. 3-6) is a disturbed clay soil with a layer of organic clay near the top containing large carbonaceous remains of roots and white pyrite outlines formed on fine rootlets. The material is from the Atchafalaya Basin of Louisiana. Print A in Fig. 3-7 shows an unretouched print made from a negative of the radiograph. The contrast between dark and light areas is not as bad as it can be. Note that one area cannot be printed

1 INCH

Fig. 3-6. Photograph of disturbed clay soil with organic material. Boring 214 UE, East
Atchafalaya Levee; depth 28 ft.

well without lightening or darkening the other area excessively. Print B shows
an improved print made with paper tissue dodging in the underexposed areas.
Print C shows a print made with regular dodging with light done by an experienced photographer.

In Fig. 3-8, print D shows the result of printing after retouching the
light areas of the negative with coccine dye, which can later be washed off
the film. This dye can be applied nonuniformly to underexposed areas in order
to achieve subtle dodging effects. Print E is a print made with an unretouched
negative using a LogEtronic printer. The LogEtronic printer scans the negative and introduces light intensity adjustments electronically to achieve a

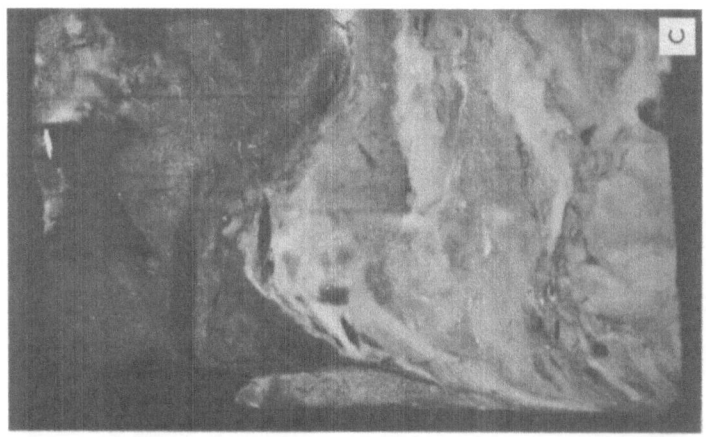

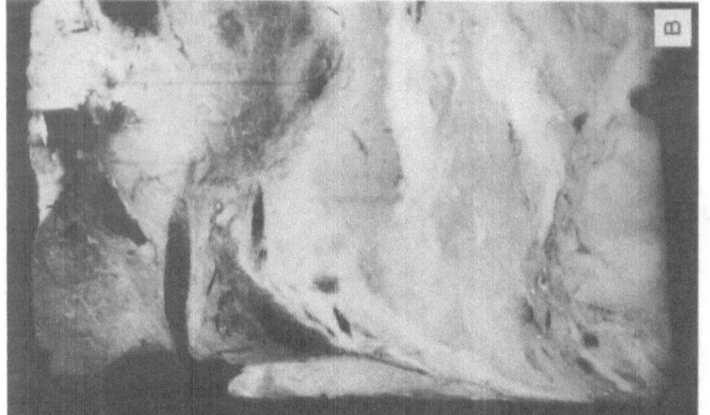

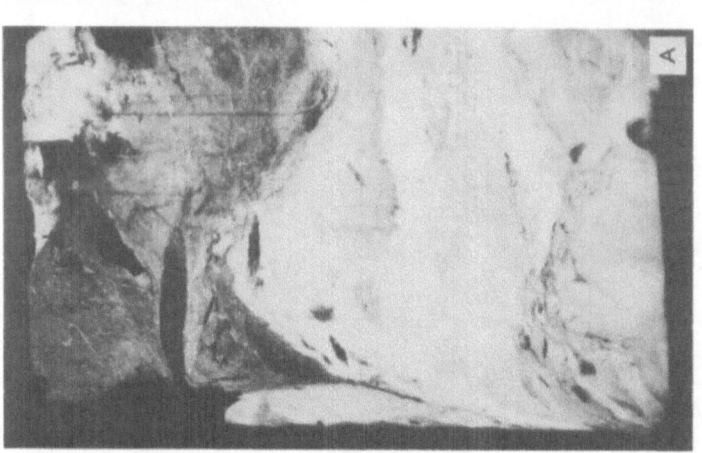

Fig. 3-7. Radiograph prints: A, unretouched; B, paper tissue dodging; C, regular light dodging.

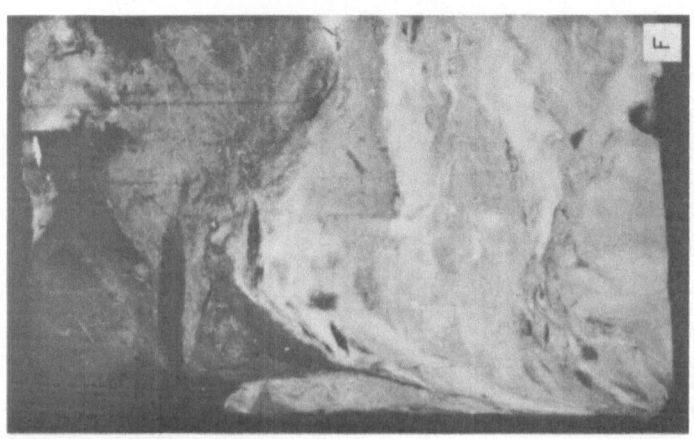

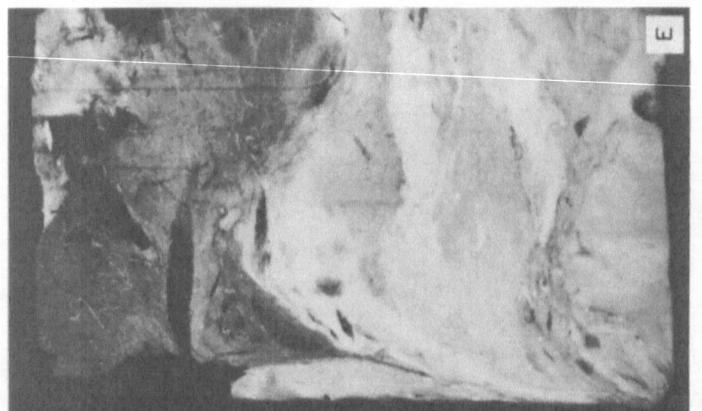

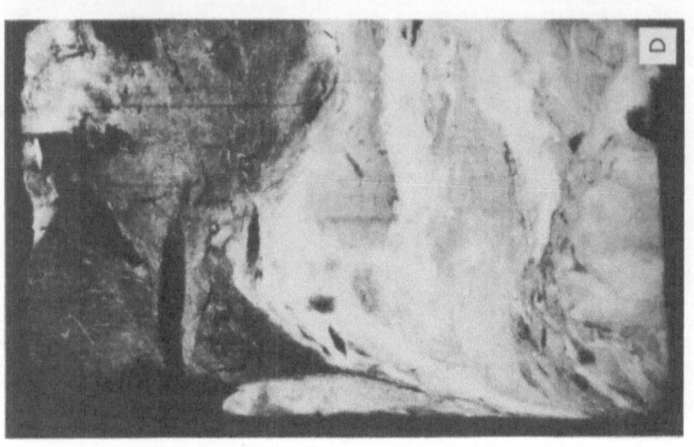

Fig. 3-8. Radiograph prints: D, coccine dye retouching; E, LogEtronic print with unretouched negative; F, LogEtronic print with coccine retouched negative.

more balanced picture density in the printing. Print F is a LogEtronic print made from negative that had been retouched with coccine dye, the best print obtainable by an automatic photoprinting method. However, a print that is on an equal level of quality can be achieved with ordinary light dodging when it is done carefully by an experienced photographer. All of the retouching methods are helpful in improving the image.

STEREORADIOGRAPHY

Stereoradiography is achieved by using paired prints, or paired portions of film, to obtain a perspective view through observation with the eyes' natural stereoscopic vision. Duplicate prints of a single radiographic image cannot give an impression of depth; it is necessary to make two exposures, either by moving the x-ray tube or by moving the sample. Also, stereoradiography is successful only if there are distinct features to be observed in the specimen, such as fossil remains, bedding convolutions, nodular material, or fracture planes (fracture planes can be missed if they lie perpendicular to the direction of radiation).

Figures 3-9 and 3-10 illustrate the possibilities of viewing a distinct object, in this case a gastropod shell. A stereographic effect is produced (Fig. 3-9) by a 5-cm lateral displacement of the core and a focal distance of 100 cm.

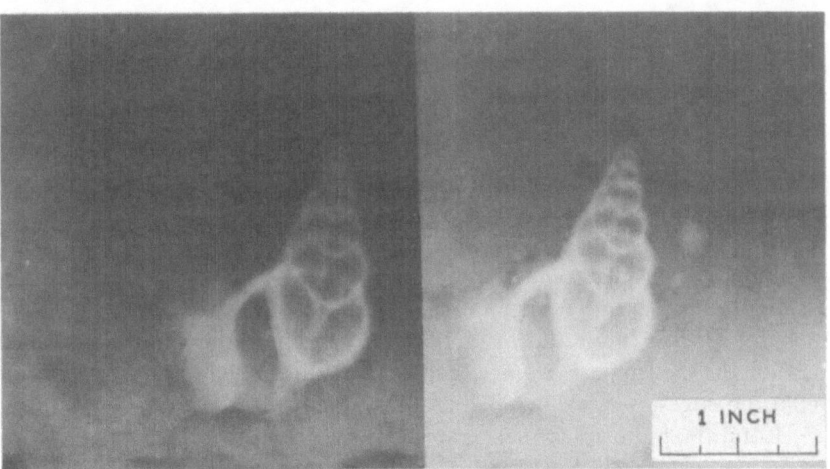

DISPLACEMENT STEREORADIOGRAPHY

Fig. 3-9. Stereoradiographic detail of a gastropod in a core, 11.5-cm diameter, from the Wilkinson Basin, Gulf of Maine. Core was displaced longitudinally 5 cm. Focal distance was 100 cm. (Courtesy Gordon S. Fraser and Adrian F. Richards.)

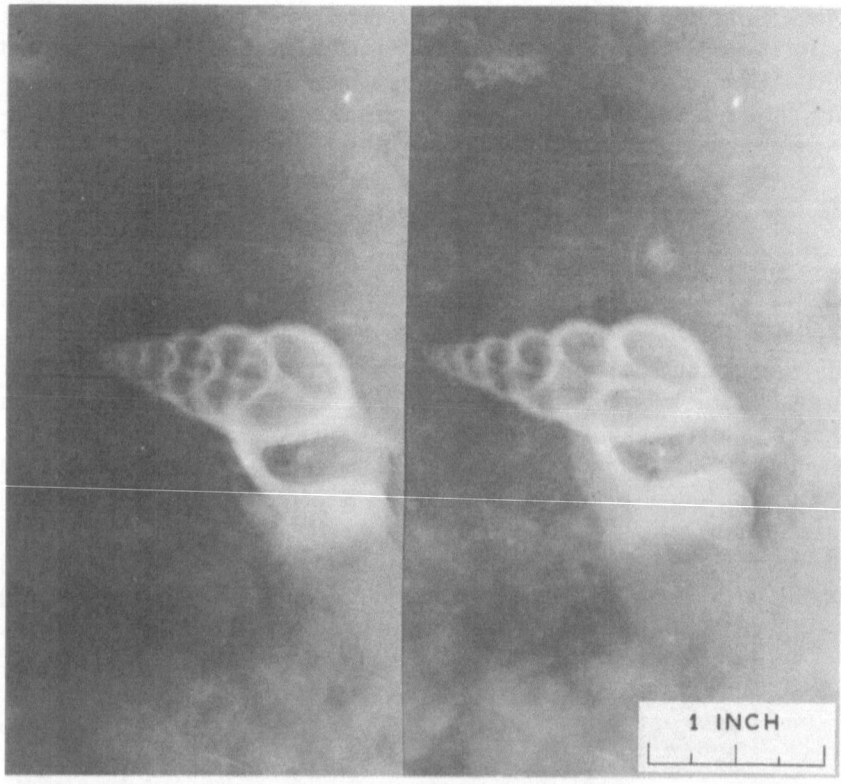

ROTATION STEREORADIOGRAPHY

Fig. 3-10. Stereoradiographic detail of the gastropod in Fig. 3-9. Pictures were taken before and after rotating the core tube 15° around its longitudinal axis. (Courtesy Gordon S. Fraser and Adrian F. Richards.)

A comparable result is obtained (Fig. 3-10) by rotating the core 15° around its longitudinal axis. The method has been described briefly by Ayer and Richards (1968), the result, in this case, a ghostly, three-dimensional effect in which the shell appears transparent and is outlined by the apparent thickening at the edges.

Rotation and displacement appear to give results of equal quality, and the method can be used, within limits, to obtain quantitative measurements. A possible application is the measurement of the progressive displacement of intersecting shear planes in soil or rock specimens undergoing compression tests. Figure 3-11 shows how such observations can be made with double exposures (also termed the parallax method). Lead reference points M_1 and

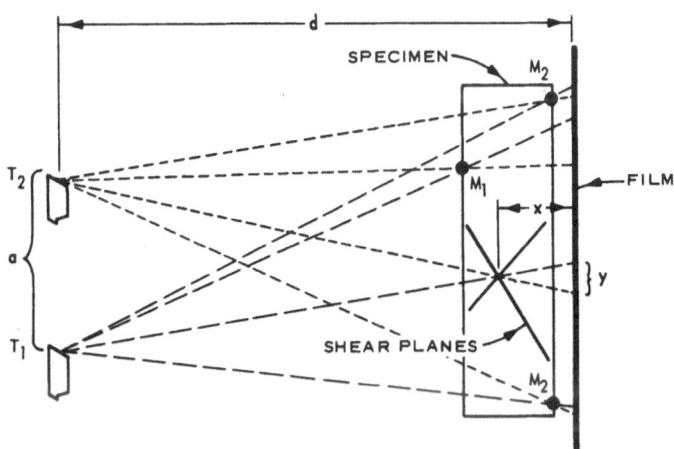

Fig. 3-11. Double exposure or parallax technique for measuring the position of failure planes in soil or rock specimens during compression testing.

M_2 are embedded in the front and back of the specimen. Exposures are made at T_1 and T_2, and the distance of movement of the x-ray tube is measured as a. The position of M_2 will change very little as a result of the tube shift, but M_1 and the shear planes will be shifted by a noticeable amount. The reference point of shear-planes intersection will be moved by the distance y. The focal distance d remains unchanged. The distance of the reference point above the film plane x can be calculated as follows:

$$x = \frac{yd}{a + y}$$

Parallax exposures can be taken at several stages during a test and measurements can be made of several reference points if they are sufficiently developed. In a remolded sample, lead shot may be buried at predetermined points in the specimen and the progressive displacement of the shot traced through the experiment. The positions of sharp, fixed points, such as lead shot, can be measured by making double exposures on one sheet of film.

A special application of stereoradiography was made in model studies of fluid motion which took place in sandstones (Rutledge, 1966). Rutledge used time-lapse sequences to analyze fluid fronts produced by injection and by gravity flow. Fluids were selected for opaqueness in radiography and for variability in viscosity. Those selected included Ditriokon (Mallinkrodt, Inc.), Baridol (Pacific Chemicals), solutions of sodium iodide, potassium iodide, and various mixtures of these substances. The studies revealed relationships between cryptostructures in sandstones and patterns of fluid motion.

Stereo viewing permitted a three-dimensional observation of the fluid front and its relation to sedimentary structures.

IMAGE INTENSIFIERS

Although film provides the sharpest radiographic imagery, there are circumstances in which instant viewing is desirable and fluorescent screens are not suitable for the purpose, such as when there is high intensity radiation and the viewer must be stationed at a distance for reasons of safety. Rapid scanning of large-diameter soil and rock cores falls in this category as does continuous observation of small scale soil models or soil testing processes using low-energy radiation with sufficient accuracy. The image produced by a fluorescent screen often can be below the threshold of brightness needed by the human eye to observe significant detail. Both groups of observations, high intensity and low intensity, can be benefited by a device known as an image intensifier.

The image intensifier is explained in the schematic diagram in Fig. 3-12. It is an evacuated tube in which radiation registers on a fluorescent screen. Photoelectrons are emitted from an attached cathode plate and are impelled at high energies, focused on another fluorescent screen on which the image detail is heightened. The image is then picked up and transmitted by an optical system and can be projected and viewed on a mirror, directed into a camera, or fed into a closed-circuit TV system and recorded on video-tape for playback. The versatility that the image intensifier makes possible recommends it for research or operations in which dynamic processes must be observed and recorded.

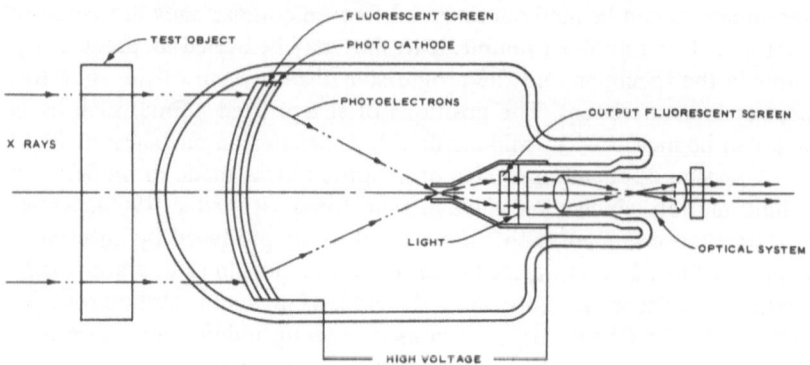

Fig. 3-12. Schematic diagram of an image intensifier.

NOTATION OF RADIOGRAPHIC DATA

The notations used in recording details observed in radiographs must ultimately depend on the investigator and the problems he must analyze. The notations described in this section are ones that the writer has found to be useful for work with unconsolidated sedimentary deposits of the lower Mississippi Valley. Examples of these notations are shown in Fig. 3-13. A routine addition of these symbols to a graphic boring log is shown in the same figure. The criteria by which these features are recognized in radiographs will be discussed in subsequent chapters of this volume.

The writer has had the symbols in Fig. 3-13 printed by the offset process in groups on wax-backed transparent sheets. The symbols may then be

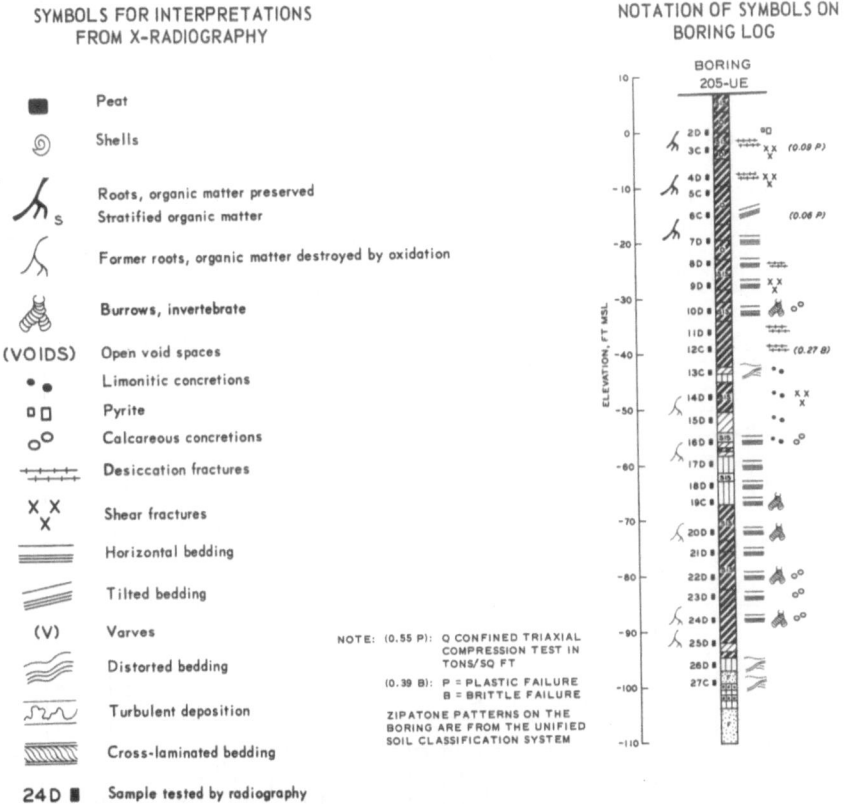

Fig. 3-13.   Examples of radiographic notations for unconsolidated sediments of the lower Mississippi Valley.

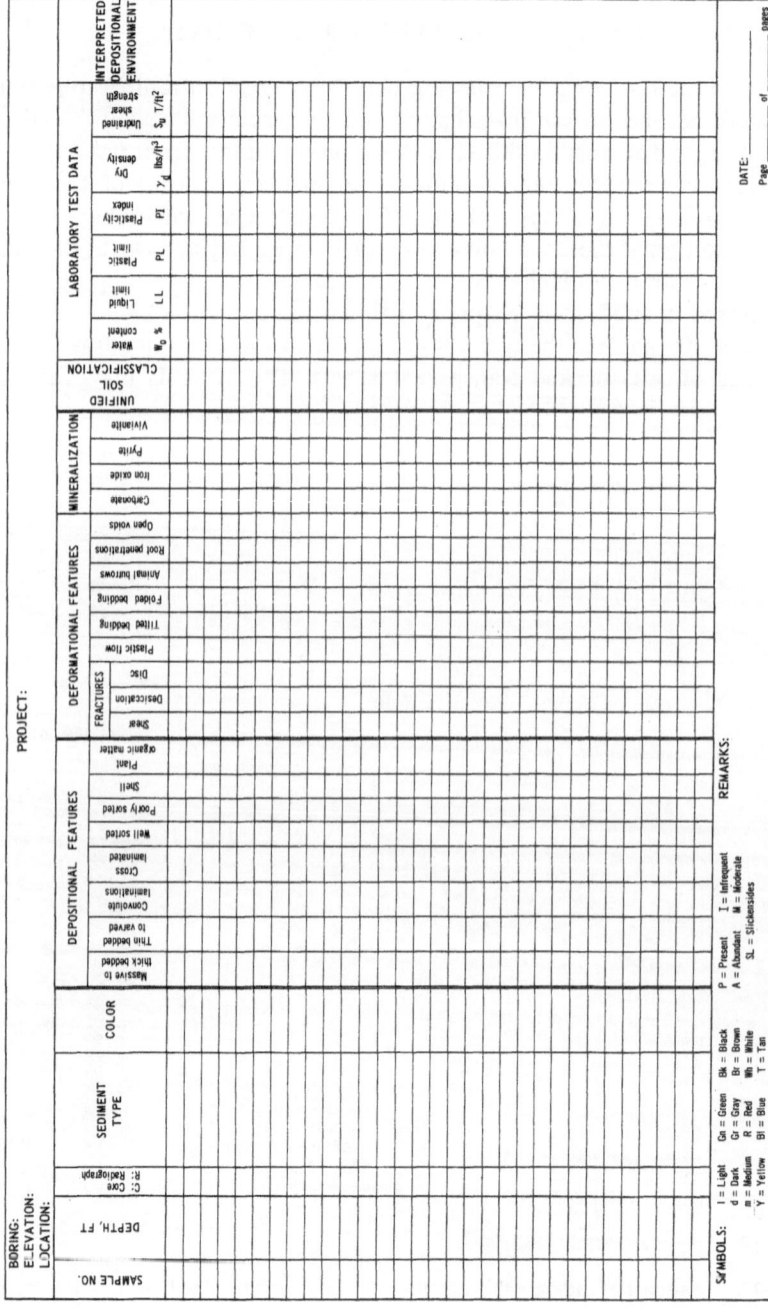

Fig. 3-14. Example of a data sheet for incorporating radiographic notes with other observations and with laboratory test data.

cut out by a draftsman and applied like Zipatone to boring logs that are being developed into a correlation section. In this way, uniform notations can be produced with a minimum of effort.

An example of a data sheet that provides both for radiographic information, other geologic observations, and test results from the soils laboratory is shown in Fig. 3-14. Such a merging of observations is a necessary part of utilizing radiographic interpretations to help arrange significant parameters for other data.

BIBLIOGRAPHY

Ayer, Nathan and Adrian F. Richards (1968). Stereoradiography of sediment cores within large-diameter plastic core tubes, *Geol. Soc. Am. Special Paper 115, Abstracts for 1967*, Boulder, Colo., pp. 462–463.

Brixner, Berlyn and J. A. Richards (1967). Projector for x-ray film, *Materials Evaluation*, pp. 31A–35A.

D'Adler-Racz, J. D. (1966). Influence of high voltage wave form and inherent filtration of x-ray generators on image quality, *Materials Evaluation*, pp. 503–506.

Eastman Kodak Company (1957). *Radiography in Modern Industry*, second edition w/supplements, Rochester, N.Y., 136 pp.

Fraser, G. S. and A. T. James (1969). Radiographic exposure guides for mud, sandstone, limestone, and shale, *Illinois State Geological Survey Circular 443*, Urbana, 19 pp.

Hamblin, W. K. (1967). Exposure charts for radiography of common rock types, *Brigham Young University Geology Studies*, Vol. 14, Provo, Utah, pp. 245–258.

Rutledge, J. R. (1966). A study of fluid migration in porous media by stereoscopic radiographic techniques, *Brigham Young University Geology Studies*, Vol. 13, Provo, Utah, pp. 89–104.

Chapter 4

LABORATORY OPERATIONS

Laboratory arrangements and operating procedures are dependent, among other things, on the types of materials, types of problems, space and funding available, and the personalities of the researchers. This volume can only treat such matters in a general way and the ideas expressed are colored by the background of the writer.

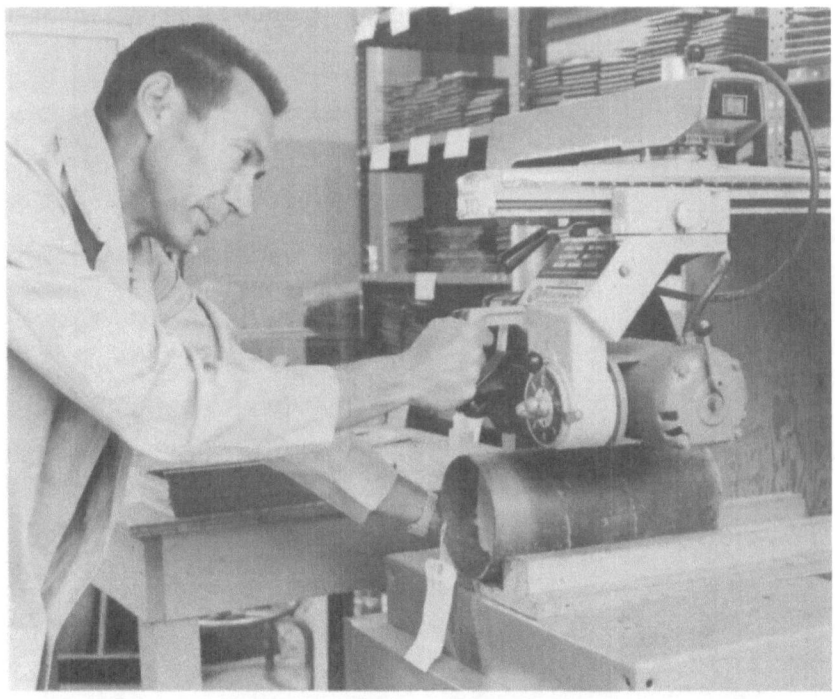

Fig. 4-1. Opening of undisturbed soil core using a radial arm saw. The core is sealed in paraffin within a cardboard cylinder.

OPERATIONAL PROCEDURES

Figure 4-1 shows an undisturbed soil core, sealed with paraffin in a cardboard cylinder, being opened with a radial arm saw. The work is being done in the Sedimentation Laboratory and in the Radiological Laboratory of the U. S. Army Engineer Waterways Experiment Station, Corps of Engineers, Vicksburg, Miss. The core is held in a wooden cradle and the saw is set to cut through the covering materials without damaging the soil. The core is then rotated 180° and another slice is taken. Then one-half of the core covering is worked off with further slices and a knife and hammer. The uncovered half of the soil core is then cut off with a piano wire using the half cylinder of wax and cardboard as a guide. The liberated core half is then put into the trimming form shown in Fig. 4-2. Using the piano wire, a slice is produced that is uniformly ⅜ in. thick; the figure shows such a slice being made from a block of remolded soil. Other forms are on hand in this laboratory that will make slices 2, 3, and 4¼ in. thick should they be desired. Thin slices produce the best pictures but thicker slices may be desired so other testing can be done

Fig. 4-2. Preparation of a thin, uniform slice of soil.

on the samples, particularly shear strength tests of undisturbed material.

The soil slice is wrapped in a sheet of resinite film, similar to Saran Wrap. This cover holds in the natural moisture. Laid on a small masonite plank, the sample is supported throughout the testing period to avoid the introduction of internal strains which would accompany bending of an unsupported sample. A large group of these slices can be accumulated and

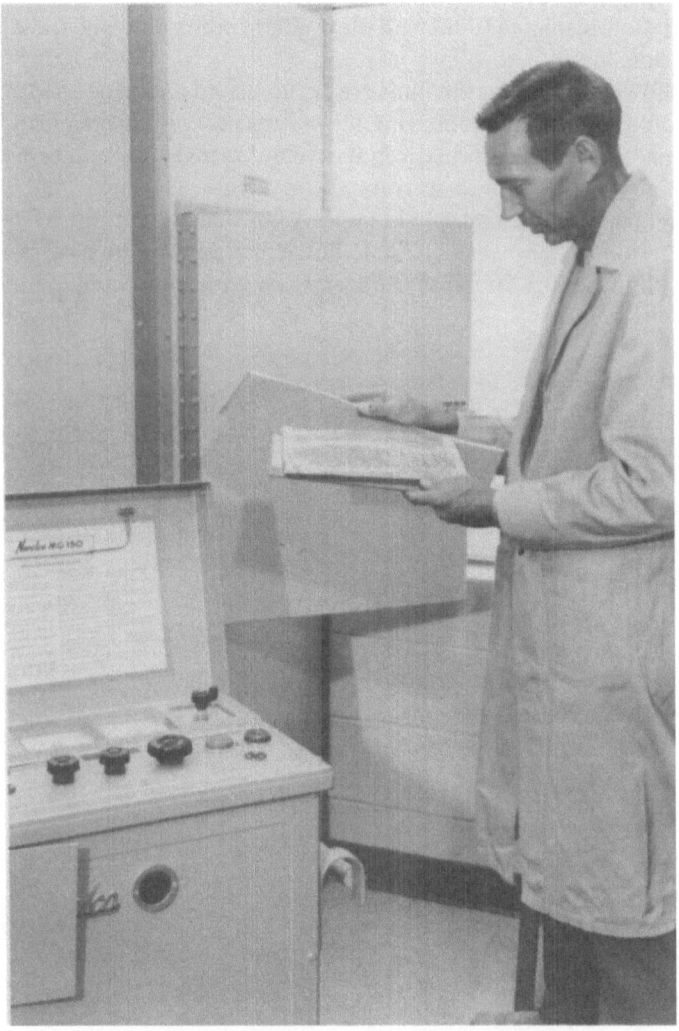

Fig. 4-3. Soil slice wrapped in resinite film is laid on x-ray film in a sealed lightproof envelope preparatory to radiation in a lead-lined cabinet.

kept in a refrigerator for a day or longer prior to radiation. They do not have to be unwrapped for radiation as the wrapping does not register in the radiograph.

Figure 4-3 shows a soil slice ready for radiation. The x-ray film can be either in a cassette loaded in a photographic darkroom or it can, as shown, be presealed in a lightproof envelope by the manufacture. Identification is maintained by placing lead numerals on the sample and maintaining a log. In this case, radiation is done in a lead-lined cabinet, but it can also be done in a radiation-insulated room with the operator controlling the radiation from the outside.

Following radiation the films are accumulated and processed as a group in a photographic darkroom using the special processing chemicals specified by the manufacturer. Automatic film developing machines can be used where the volume is sufficient to warrant them. Figure 4-4 shows the developed x-ray film being studied on a light table. A film density of 1 to 1.5 is handy for observation with any sort of light, but it is desirable to have a variable-intensity light source for denser film which should show detail in film as dark as 4.5.

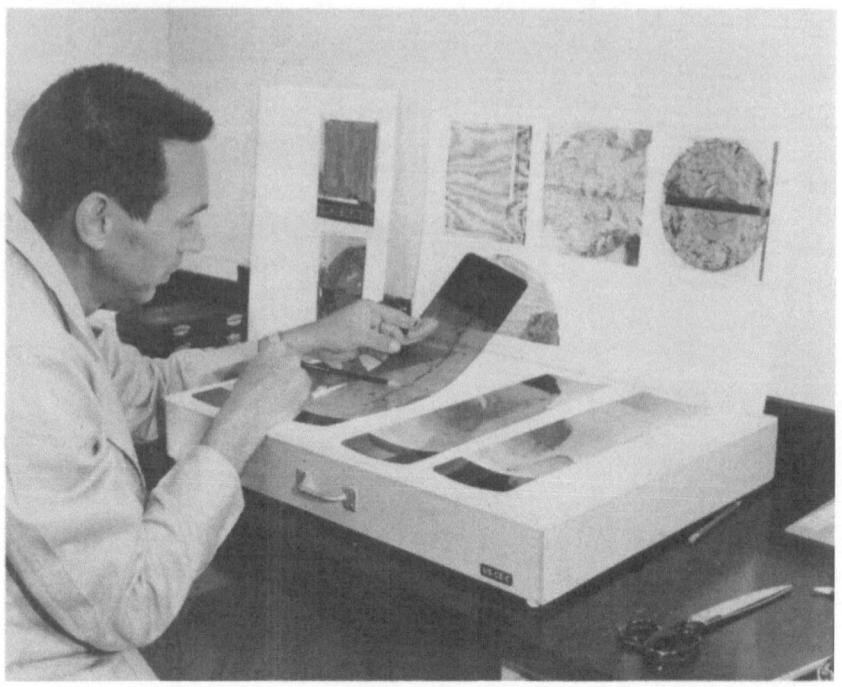

Fig. 4-4.   Developed x-ray film being observed on a light table.

Material such as clay shale that is too stiff to be trimmed with piano wire, yet is not so rigid as to require a rock saw, can be cut with a band saw provided with a flat sliding surface, permitting slices to be taken. For hard rock, the cutting must be done with a rock saw.

Often, soft materials may contain tree roots, concretionary nodules, shells, or isolated pieces of gravel. Roots can be cut with a hacksaw, but with nodular materials and large shells, it is necessary to cut around them, or pluck them out, finish the slice, and set the plucked material back in. The unnatural disturbance that is introduced can be allowed for in the interpretations.

Radiographic observation of completely noncohesive material, such as loose sand, can be done if the sand is impregnated with nonsetting gels, oil, or thin clay slurries sufficient to permit handling. For specimens that are not formed into slices but need to be radiographed as cylinders, nodules, or other forms, compensations are desirable to obtain samples of uniform thickness and uniform density for passage of the x rays. Such objects can be set in beds of uniform, fine sand, or fitted in plaster-of-paris molds.

Though the procedures just described involve only static observations, they are adaptable to many laboratory testing experiments. Test specimens subjected to experimentation can be observed before testing and after testing. Blocks may be cut out of large models subjected to deformations induced by such tests as blast effects, loading on footings, or disturbances from wheel tracking, and observed in detail for deformation with control-sample comparisons.

CONTINUOUS SCANNING

Continuous scanning operations in the laboratory may be considered under the two categories of low-intensity and high-intensity radiation. These systems, discussed in the following paragraphs, may or may not be in use as described, but they incorporate features that individually have been tested and are known to be practical.

Low-Intensity Radiation

Using an image intensifier, radiation can be kept at a level sufficiently low for the x-ray tube to be taken out of the shielded room or the lead-lined cabinet allowing operations completely in the open or behind a lead-glass plate. The operator can be protected by a lead-vinyl apron, lead-vinyl gloves, and lead-glass goggles.

It is presumed that such a setup would be useful for model studies involving either continuous deformation or deformation by discontinuous loading. Possible problems include movement along residual shear planes,

partial liquefaction of sandy soils through pore pressure changes, and liquefaction of sands by simulations of earthquake shocks. Mechanisms of
deformation by expansive clays can also be examined by this means, and other
investigations involving the effects of naturally occurring geologic features
on the performance of soils tests.

A schematic diagram for a possible operating unit is shown in Fig. 4-5.
The unit shows a low-intensity x-ray source, collimators to control the spread
of radiation so that it is accessible only to the model, the model, an image
intensifier to heighten the low-intensity radiation, and appropriate recording
devices.

For best registration, models should be contained between rigid, parallel
plates of glass or Plexiglas. In order to reduce friction, the plates can be lined
with flexible membranes, if the resulting increase in lateral compressibility is

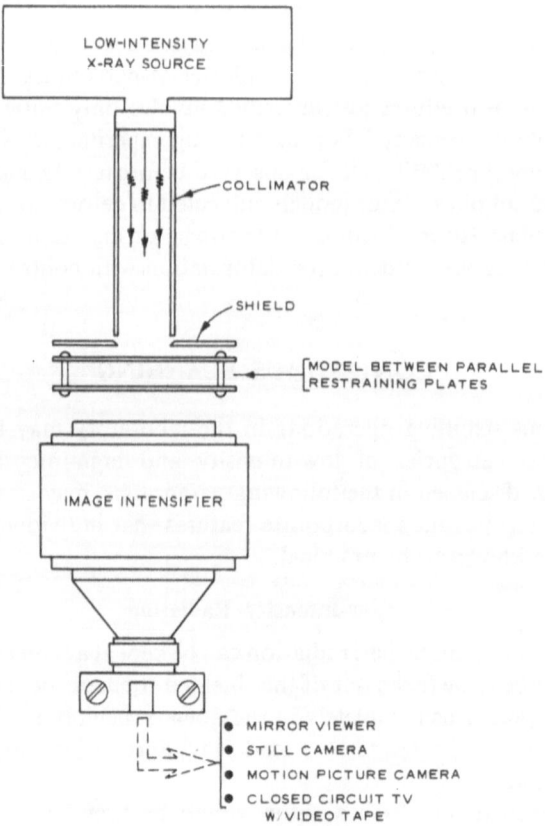

Fig. 4-5. Schematic diagram of possible operating unit for continuous scanning with low-
intensity radiation.

compatible with the type of test being conducted and the contact surfaces lubricated with silicone grease. Where uniform thicknesses are maintained, quantification of density changes is feasible.

Models not contained between parallel plates can be studied if compensations are provided for irregularities in shape.

Registration may be through a mirror viewer connected to the image intensifier, or a still or motion-picture camera used in combination with the mirror. Closed circuit TV may be installed that connects with the optics of the image intensifier. The advantage of using closed circuit TV is that it may contain a video tape unit for recording and playback.

High-Intensity Radiation

A system of continuous radiation at high intensity presumes that the equipment is housed in a well-shielded area or in an isolated and shielded

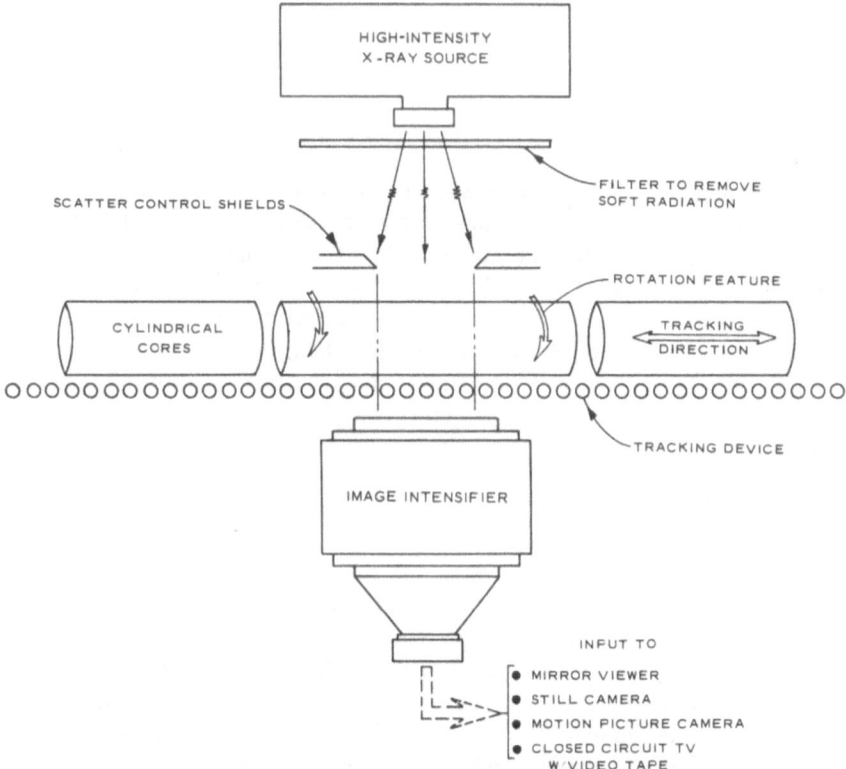

Fig. 4-6. Schematic diagram of possible operating unit for continuous scanning with high-intensity radiation and controlled tracking.

room. Such a system complete with the necessary tracking device, would be useful for rapid scanning of soil and rock cores, soil cores being evaluated for variations in depositional environment, evidence of structural deformation, and presence or absence of critical layers. The data can be used both for correlation purposes and for the assignment of laboratory testing.

A schematic diagram of a possible operating unit is shown in Fig. 4-6. High-intensity radiation is directed downward for safety purposes. Filters should be used to remove soft radiation and shields are desirable in order to minimize scatter. The tracking device should be loadable outside the radiation area.

Most materials handled are cylindrical cores and might include samples of unconsolidated or noncohesive materials taken in steel tubes but it probably is not practical to build in compensations for cylindrical shapes; it is sufficient to look only at the center portions of the cores. However, the cores can be rotated as part of the inspection thereby revealing what may have been missed initially along an edge.

Registration is made with the use of an image intensifier and with a selection of image recording units.

FLASH RADIOGRAPHY

A flash x-ray source generates a series of high-intensity radiation bursts for extremely short periods of time—microseconds or nanoseconds; output is as great as 2500 A at 330 kV. Figure 4-7 shows a schematic arrangement for a flash system. Its advantage is that it can record dynamic changes in soils where blast loadings are applied and it can record the propagation of stress wave fronts in soil by registering soil density changes. Samples can be studied with thicknesses to about 5 in., and it can also trace the motion of pellets buried in soil. Its disadvantage is that it is expensive. The technique and its applications are discussed further in Chapter 9.

GAMMA RADIATION

Gamma radiation, from radioactive source materials, produces radiographs in the same manner as x rays but with the disadvantage of being extremely slow. Its prime advantage is in economy. A working quantity of cobalt 60 may be obtained for only a few hundred dollars compared to many thousands of dollars for a functional x-ray system. Radioactive isotopes may also have special applications in some model tests in soil mechanics because of the greater freedom with which an isotope can be placed either adjacent to, or within the model.

In order to improve contrast and reduce exposure time, image intensify-

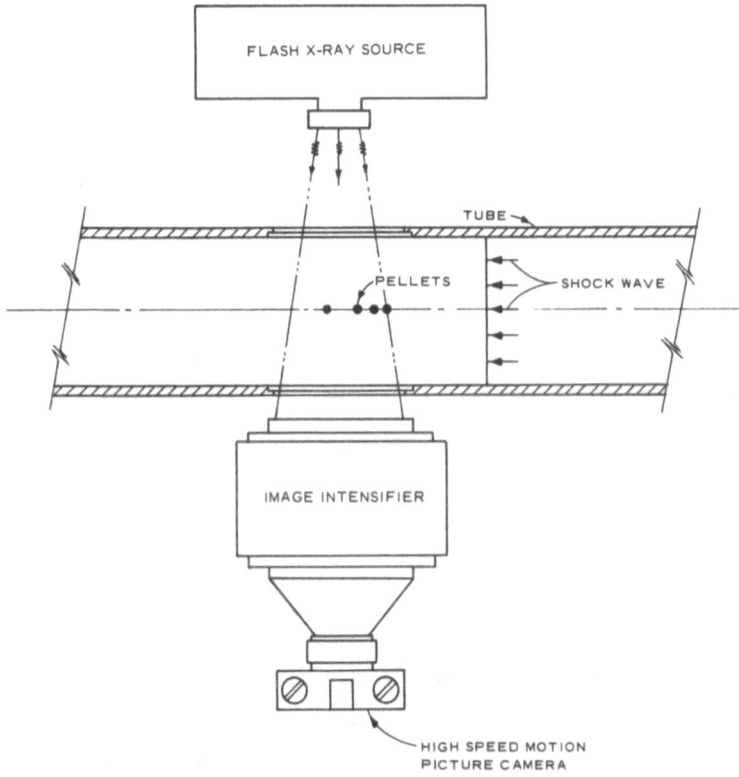

Fig. 4-7. Schematic diagram of a flash x-ray system for observing dynamic soil phenomena (after Bloedow, 1962).

ing screens are generally used and either lead-foil or calcium tungstate screens should be placed in the film cassette. The effectiveness of intensifying screens varies widely with changes in radiation strength, and most exposure values must be obtained by trial and error. However, Reed suggests the film speeds given in Table 4-1 as useful with intensifying screens. Exposure

Table 4-1

Kodak x-ray film	Film speed	Contrast
Type M w/lead foil	30	High
Type A w/lead foil	100	High
Type F w/calcium tungstate	360	Low
Type K w/lead foil	500	Medium

time may be reduced further by developing the film for eight minutes instead of the usual five.

Weiss (1966) showed that a very small gamma-ray unit could be made to function effectively by using an iodine-125 source of less than 5 millicuries. Because of the low intensity of radiation, intensifying screens could not be used and exposures had to be made at short focal distances, 12.5 cm and 25 cm. Consequently, there was geometric distortion of the images. However, for qualitative work, this was of no concern. His experiment revealed details of sedimentation patterns and other features in slices of rock.

SAFETY

Radiation safety is a special field in itself. Though radiation can be dangerous, it can also be handled effectively with almost no risk, provided proper design measures are taken. The earth scientist who begins working with radiation would do best to inform himself fully on safety requirements.

The manufacturers of x-ray equipment are helpful in suggesting designs that are safe and most instruction courses for technicians emphasize necessary safety precautions. Excellent guides and training for workers in the United States are provided by the U.S. Public Health Service. Information on their safety training programs may be obtained by contacting the Department of Health, Education, and Welfare, Public Health Service, National Cen-

Table 4-2

	Average weekly dose,* rem	Maximum 13-week dose, rem	Maximum yearly dose, rem	Maximum accumulated dose,† rem
Controlled areas				
Whole body, gonads, blood-forming organs, and lens of eye	0.1	3	—	5(N–18)‡
Skin of whole body	—	10	30	—
Hands and forearms, head, neck, feet, and ankles	—	25	75	—
Environs				
Any part of body	0.1	—	0.5	—

* The dose equivalent in rems may be assumed to be equal to the exposure in roentgens. The roentgen is the quantity of x or gamma radiation such that the associated emission per 0.001293 g of dry air—equal to 1 cm³ at 0°C and 760 mm Hg—produces, in air, ions carrying 1 electrostatic unit of electricity of either sign. Exposure of patients for medical and dental purposes is not included in the maximum permissible dose equivalent.
† When the previous occupational exposure history of an individual is not definitely known, it shall be assumed that he has already received the full dose permitted by the formula 5(N− 18). Persons who were exposed in accordance with the former maximum permissible weekly dose of 0.3 rem and who have accumulated a dose higher than that permitted by the formula shall be restricted to a maximum yearly dose of 5 rem.
‡ Age in years and is greater than 18.

ter for Radiological Health Training and Manpower Development Program, Rockville, Maryland 20852. The Department maintains regional instruction centers in nine cities in the United States.

As a general practice, casual personnel, women known to be pregnant, and individuals under 18 years of age should not be permitted in radiation areas.

Persons working with radiation should be provided with film badges and equipment should be checked periodically with radiation measuring devices. Maximum permissible radiation dosages are specified by the Atomic Energy Commission (1966) and the National Bureau of Standards (1966) as well as by numerous other agencies. Guides to radiation safety and radiation detection are provided in numerous publications (see NCRP Report

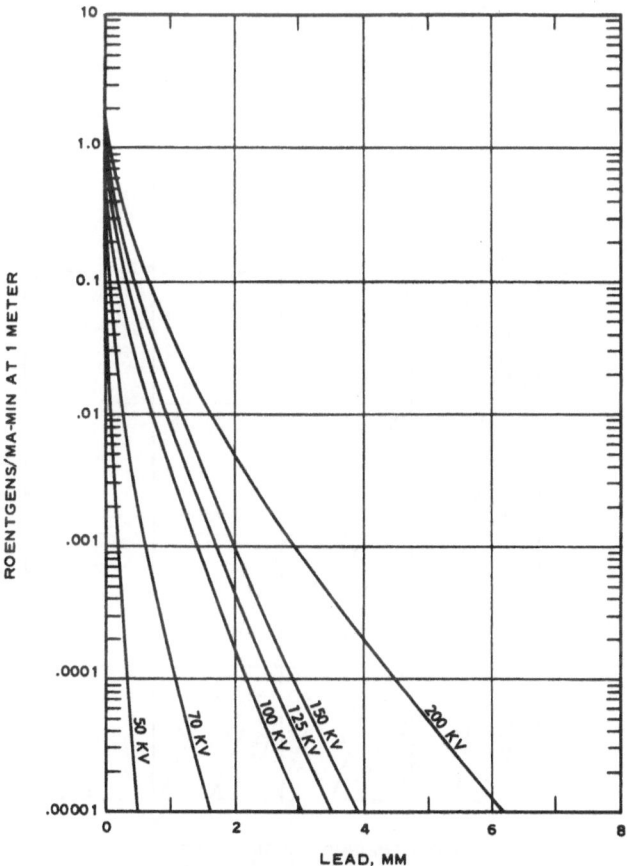

Fig. 4-8. Attentuation in lead of x rays by peak potentials of 50 to 200 kV (after National Bureau of Standards, 1966).

33; Price, 1958; Heffan, 1967; Morgan & Corrigan, 1955). Specifications used in general practice often are not in agreement, as some organizations may wish to require more stringent safety practices than are absolutely necessary in order to work well within the limits of safety. Permissible dose equivalents indicated by the Bureau of Standards are given in Table 4-2.

Designs for lead and concrete shielding are shown in Figs. 4-8 and 4-9, respectively. Thicknesses are those that provide for full absorption of the x-ray beam 1m from the source and in a plane that is perpendicular to the beam. It is good practice to overdesign shielding requirements. Figure 4-8 and 4-9 show only the minimum values that may be used.

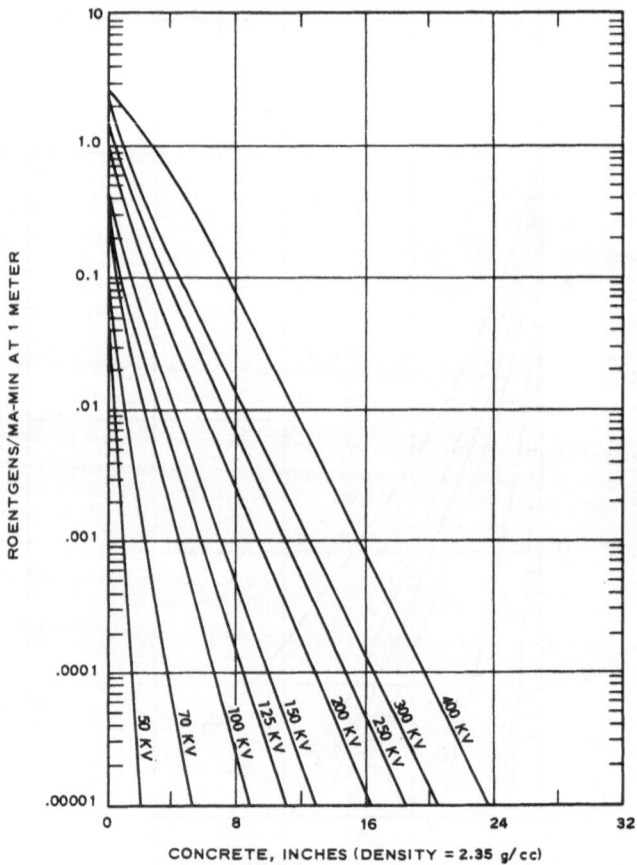

Fig. 4-9. Attentuation in concrete of x rays produced by peak potentials of 50 to 400 kV (after National Bureau of Standards, 1966).

BIBLIOGRAPHY

Bloedow, F. H. (1962). *Radiographic Instrumentation Study*, Research Directorate, Air Force Special Weapons Center, Kirtland AFB, New Mexico, 143 pp.

Heffan, Howard (1967). Radiation safety in radiography, *Materials Evaluation*, pp. 83–90.

Morgan, Russell H. and K. E. Corrigan (eds, 1955). *Handbook of Radiology*, section 6: radiation protection, Yearbook Publishers, Chicago, pp. 419–448.

National Bureau of Standards (1964, revised 1966). *Handbook 93: Safety Standard for Non-Medical X-Ray and Sealed Gamma-Ray Sources*, part 1, General, Department of Commerce, Washington, D. C., 60 pp.

National Council on Radiation Protection and Measurements Report No. 33 (1968). *Medical X-Ray and Gamma-Ray Protection for Energies up to 10 MeV*, Washington, 66 pp.

Price, William J. (1958). *Nuclear Radiation Detection*, second edition, McGraw-Hill, New York, 430 pp.

Reed, Malcolm E. *Cobalt-60 Radiography in Industry*, Tracerlab, Waltham, Mass. 29 pp.

United States Atomic Energy Commission (1966). *Rules and Regulations*, Title 10—Atomic Energy, Part 20—Standards for protection against radiation, Washington, D. C., pp. 63–78.

Weiss, Malcolm P. (1966). X-radiography of rocks with I-125 source, *Bull. AAPG* **50** (7): 1507–1510.

Chapter 5

SEDIMENTATION STUDIES

The first effective application of radiography in the study of sediments was described by Hamblin (1962). His work revealed that so-called "thick bedded" or "massive" sedimentary deposits really contain many complex primary structures that are visible with x rays but are otherwise either poorly expressed or invisible. His observations were quickly confirmed by Calvert and Veevers (1962) and extended to unconsolidated marine sediments. They reported that the method:

1. Confirmed the homogeneity of sediments that appear homogeneous in reflected light.
2. Resolved fine details that are partially evident in reflected light.
3. Revealed particles, such as shells or nodules, that lie beneath the surface of a slice.
4. Revealed structures that are not visible in reflected light.

Other papers referring to the technique were published by Rioult and Riby (1963), Bouma (1963; 1964a, b, c; 1968), Bouma and Marshall (1964), Bouma and Shepard (1964), Coleman (1966), Coleman and Ho (1967), Kolb and Kaufman (1967), Krinitzsky (1966, 1967, 1970), Krinitzsky and Smith (1969), and Shepard, Dill, and von Rad (1969).

Bouma (1964c) provided an exceptionally well illustrated review of the primary and secondary structures revealed by x radiography in marine sediments. He pointed out that the effects of burrowing and slumping are brought out and that radiography should be applied before samples are selected for other investigations. He also cautioned that the technique should be considered as a supplement to other methods of study but not as a replacement.

Coleman (1966) showed that the features revealed by radiography are useful for dividing sedimentary deposits into subenvironments of deposition. The applicability of the technique for correlations with engineering properties and for correlations between borings was shown by Krinitzsky (1966, 1967, and 1969).

The applicability of radiography for defining complex structural deformations in deeply buried clays of the Mississippi River Delta was shown by Kolb and Kaufman (1967).

ALLUVIAL AND DELTAIC ENVIRONMENTS OF DEPOSITION

Sedimentation processes in the alluvial valley and deltaic plain of the lower Mississippi River can be considered as representative of similar fluvial deltaic and coastal deposition of sediments in other parts of the world. The general distribution of these sediments is shown in Fig. 5-1. Note that the valley is nearly 600 miles long from Cairo, Illinois, to the Gulf of Mexico, about 30 miles wide in its most constricted part, about 75 miles southwest of Vicksburg, and about 125 miles wide at its widest, south of Memphis. The entire area of the alluvial valley was entrenched by river erosion during stages of the Pleistocene when sea level was as much as 450 ft lower than it is now. During intervening times, the valley received sedimentation. Part of

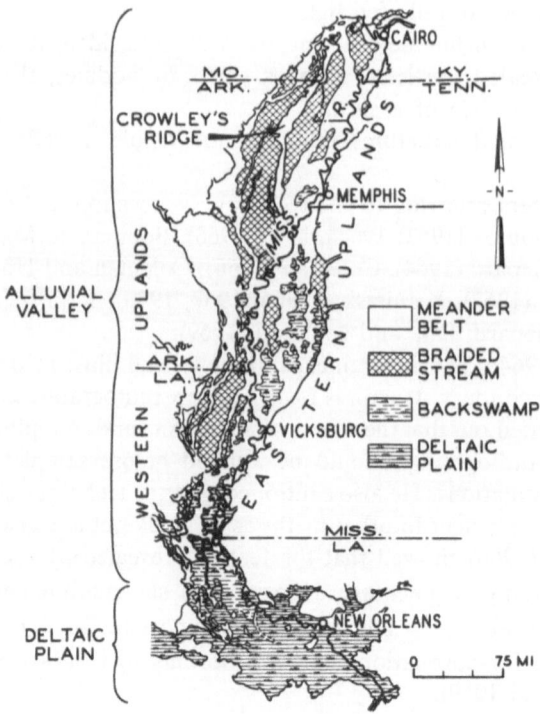

Fig. 5-1. Distribution of sediments in the lower Mississippi Valley.

the sedimentary fill is braided stream deposits laid down as glacial outwash, since then extensively reworked by meandering of both the Mississippi River and its tributaries. The deltaic plain is composed of a series of deltaic masses and corresponding coastwise deposition and reworking of sediments. Deposition in the delta was accompanied by almost continuous settlement.

The major sedimentary deposits laid down in the alluvial valley and deltaic plain are described briefly in Table 5-1, classified according to manner of deposition. These depositional environments are useful for interpreting and mapping individual deposits that incorporate similarities in composition, engineering properties, chronology, and other characteristics. Even the deposits of smallest extent, such as channel fill deposits, are large enough that further subdivision would be useful in order to trace details of their strength characteristics or the continuity of correlation horizons.

A large-scale example showing the need for greater detail in interpretations is the backswamp of the Atchafalaya Basin. The Atchafalaya Basin lies in the western part of the Mississippi alluvial valley in Louisiana. It is about 100 miles long and about 30 miles in maximum width and is continuously filled with backswamp deposits to a depth of 100–150 ft. Figure 5-2 shows a schematic section through a part of the basin. The basin is bordered by a levee system built on these backswamp deposits, which must be a record for extent of treacherous foundation conditions. There are approximately 200 miles of levees and they have settled an average of 5–10 ft over their entire length. One point along the levee has settled a measured vertical distance of 29 ft. How, then, may the foundation soils be investigated if it is one single depositional unit?

The difficulty with usual modes of observation is that the Atchafalaya backswamp deposits look very much the same everywhere. They are mostly fat clays, and to a visual inspection they offer very little in the way of distinguishing characteristics. Even the presence or absence of organic matter is largely masked by the prevailing matrix of fat clay. Radiocarbon age determinations made on organic matter collected from Atchafalaya levee borings showed that the backswamp deposits have been forming for somewhat longer than 15,000 years and were developing continuously to the present.

It was with a problem such as this that the writer experimented with possible applications of radiography. Approximately 1000 undisturbed soil cores were collected from 42 borings made along the Atchafalaya levees. The borings had full or nearly full penetrations of the backswamp, and were each 5 in. in diameter and 1 ft in length. All cores were opened and in addition to observations on soil type, stratification, fractures, organic matter, fossils, burrows, secondary mineralization, etc., were prepared routinely for radiography.

The radiographic samples were prepared as 5-mm-thick, core-length

Table 5-1. Major Sedimentary Deposits in the Lower Mississippi Valley

TOPSTRATUM: Complex, cohesive, mostly fine-grained sediments comprising the upper portion of the alluvial fill.	Calcareous loess: Uniform calcareous clayey silt, thickness 10 to 100 ft on hill crests. Believed to be eolian and derived from present and former braided stream deposits in the Valley. Source of clayey silt forming alluvial apron.

BRAIDED STREAM DEPOSITS:
Complex interlayered clays, silts, and fine sands. Thickness approximately 40 ft.
Apron deposits: Alluvial aprons fronting the valley walls. Composed of layers of redeposited loessial clayey silts, loam, and coarser detritus. Up to 25 ft or more thick, up to 4 or 5 miles wide.

BACKSWAMP DEPOSITS:
Clays to silty clays. Extensive clayey layers developed in low-lying areas adjacent to meander belts. From 35 ft thick in the northern part of valley to 120 ft thick in south. Accumulations of finest materials brought in by floodwaters.

MEANDER BELT DEPOSITS:
Natural levee: Silty clays and silty sands. Thin blanketing veneer, 20 ft or less, overlying other deposits. Forms natural dikes along banks of rivers. Widths of 1/2 to 1-1/2 miles. Includes crevasse deposits which result from breaching of natural levee during flood. Crevasses typically form dendritic patterns of sedimentary veneers several square miles in areal extent.
Abandoned channel (clay plugs): Highly plastic clays, silty clays, filling oxbow lakes produced by cutoffs of meander loops. Clayey masses 70 to 120 ft thick. Silty in arms of abandoned loops.
Abandoned course: Sandy silts, clayey silts filling abandoned river courses. Depths to 90 ft.
Point bar, upper portion: Sands, with silty sands and some clay. Complex internal patterns of deposition. Sediments laid down in growing point bars; groups of point bars, containing numerous swales, may be many miles in extent. Thickness approximately 35 ft. Includes swales: silts, clayey silts, sandy silts which fill lows between bars. Long, narrow, arcuate deposits often several miles in length, a few hundred feet in maximum width, with depths up to 60 ft.

SUBSTRATUM: Noncohesive, coarse- to medium-grained sediments comprising the lower portion of the alluvial fill.

MEANDER BELT DEPOSITS:
Point bar, lower portion: Sands, sometimes containing gravels. Laid down in growing point bars.

SANDS, SANDS AND GRAVELS:
Clean sands, also sands with chert gravels in varying quantities. Sediments deposited prior to development of Recent meander belts. May be Pleistocene to Early Recent in age. Maximum thicknesses to 200 ft.

ALLUVIAL VALLEY

DELTAIC PLAIN

A small modern bird's-foot delta at the Mississippi River passes plus six older and larger deltas, interleaved with each other, plus deposits of intervening areas of lakes and bays, swamps, beaches, and ancient stream courses. Sediments are 150 to 700 ft in total thickness and are predominantly clays. The western part is underlain by coarse-grained substratum deposits left by an ancestral Mississippi River course when sea level was lower than at present.

Prodelta: Homogeneous clays deposited offshore; thickness 50 to 600 ft.

Intradelta: Intricately interfingered clays, silts, and sands; coarse portion of subaqueous delta formed around major courses and distributaries. Thickness associated with present Mississippi River course, 200 ft; with older courses 25 to 100 ft.

Interdistributary: Clays with minor amounts of silt, sand, and organic matter; thinly layered, "varved" appearance. Deposited between major distributaries. Thicknesses similar to intradelta.

Marsh: Watery organic ooze to firm organic silts and clays. Forms on land surface of Deltaic Plain. Maximum thickness 30+ ft, average 10 ft.

Swamp: Organic clays with silts and sands. Forms in inner borders of marsh subject to freshwater inundation. Thickness 3 to 10 ft.

Tidal channel: Organic clay to fat clay filling abandoned tidal channels. Cross section 200 ft wide, 25 ft deep.

Bay-sound: Sandy, silty clays, poorly stratified and mottled; thickness 3 to 20 ft, average 15 ft.

Nearshore Gulf: Sands with thin layers of shells, silt, clay; deposited in open ocean beyond barrier beaches. Thickness to 25 ft.

Point bar: Same as corresponding deposits in the Alluvial Valley but restricted to the more prominent bends of present and abandoned river courses; depths 100+ ft. Bedded fine-grained topstratum 25 to 75 ft; clean sand substratum. Some deposits consist almost entirely of silt.

Natural levee: Narrow bands, 1/4 to 1-1/2 miles wide, bordering present and abandoned river courses and distributaries. Layered fat and lean clays and sandy silt. Thickness to 25 ft along rivers; 5 ft or less along distributaries.

Estuary: Sandy facies correlative with nearshore Gulf deposits but filling minor valleys entrenched into the underlying Pleistocene surface.

Lake: Poorly stratified clays with silt, sand, and shell; 2 to 25 ft thick.

Sand beach: Developed in areas bordering the open Gulf where the delta is nonactive. One mile wide to 10 miles long; average 30 ft thick. Predominantly sand with some silt and shells.

Shell beach: Developed on landward shores of bays, sounds, and marshland lakes; 25 to 200 ft wide and 1-1/2 to 6 ft high. Chiefly shell material containing silt and sand.

Reef: Shell clusters developed principally in bay-sound areas. Tens of feet to 1/2 mile wide, tens of miles in length. Shell thicknesses average 5 to 10 ft.

Abandoned course: Abandoned Mississippi River courses, 1/2 mile average width, 75 to 150 ft deep. Sandy material in lower portion, silts and clays above.

Abandoned distributary: Former distributary channel up to 1000 ft wide, 10 to 55+ ft deep filled with a wedge of fine sand to silty clay at the upstream end and which grades to clayey sediments.

FORMER DELTAIC PLAIN

Pleistocene clays: Clays of former (pre-Recent) Deltaic Plain of the Mississippi River which underlies the Recent deposits. Laid down under environments similar to those in the Recent Deltaic Plain. Clays were formerly subjected to consolidation, subaerial erosion, desiccation, oxidation, and, subsequently, reduction by groundwaters.

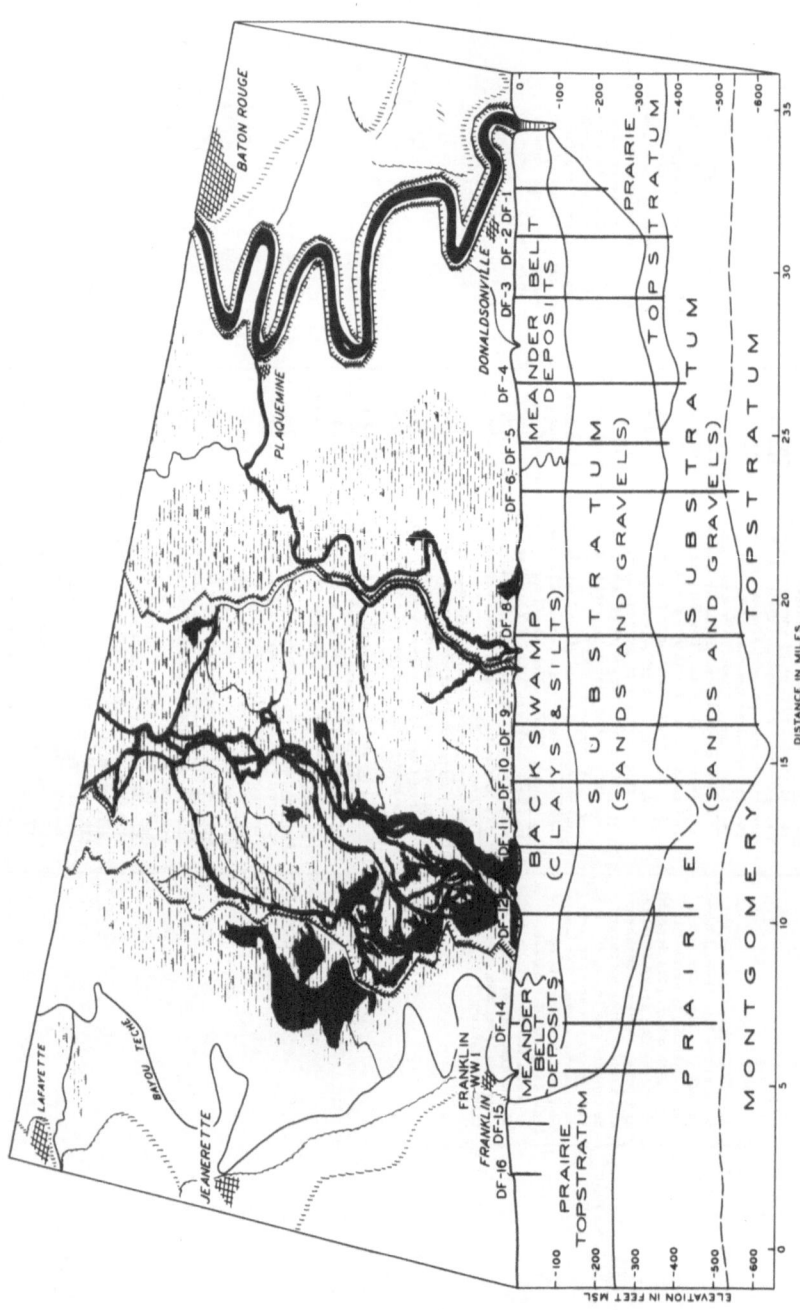

Fig. 5-2. General subsurface distribution of sediments in the Atchafalaya Basin of the lower Mississippi Valley.

slices from the centers of the cores by cutting with piano wire. Radiography was done with a Philips Industrial x-ray unit using a 100-kV beryllium-window tube, and the records were made on Kodak Type M Industrial x-ray film. Exposure was for 45 sec at 18 mA and 25 kV with a focal distance of 18 in. An exposure of 20 sec was used with highly organic materials, such as peat.

Study of the radiographs obtained showed that the backswamp was divisible into several distinct environments of deposition, chiefly, those of lakes, sometimes with thin deltaic facies where rivers have entered the lakes, swamps that were well drained during their development, and swamps that were poorly drained. The identifying features of these environments and the radiographic characteristics that are associated with them are discussed in the following paragraphs.

LAKE ENVIRONMENT

Lake environments in the Atchafalaya Basin were those of shallow bodies of fresh water, ecological evidence suggesting that water depths were never greater than 18 ft and were generally much shallower (*cf.* Coleman, 1966).

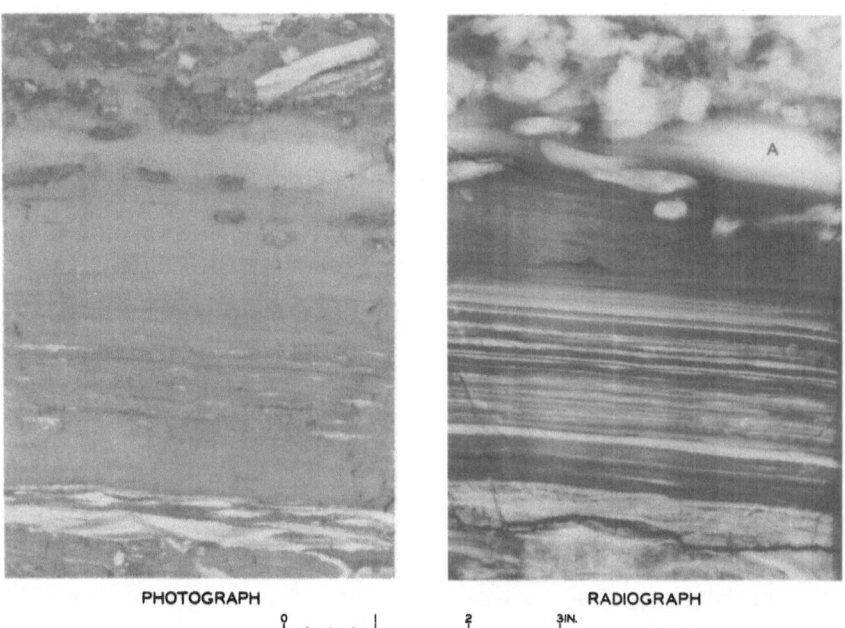

PHOTOGRAPH   RADIOGRAPH

Fig. 5-3. Lake deposit containing varved clays and effects of burrowing animals. An incipient calcareous concretion is shown at A. East Atchafalaya Levee Boring 202; depth 60 ft.

The sediments received were largely clays with much lesser amounts of silt. Occasionally there were introductions of sands of fine- to medium-grain sizes. These are considerably localized and are believed to be related to deltaic facies and channel fill in streams bringing flood detritus into the basin. Current action and winnowing processes have spread thin layers of such sands over portions of the bottoms of the larger lakes. The clays are generally gray to dark gray and indicate accumulation in a nonoxidizing environment. However, some reddish clays are found in upper parts of the section and are believed to have been brought in with reddish coloration by the Red River.

The predominating radiographic features of lake facies in the Atchafalaya backswamp deposits are:

Stratification. Stratification is generally well developed. Even thick masses of fat clay, laid down during quiet, continuous deposition, have delicate indications of layering that show up with radiography. A striking example of such layering is shown in Fig. 5-3. Note the delicate laminations that show up in the radiograph. From the photograph, it is seen that part is due to textural changes in the material, however, the photograph does not bring out the full delicacy of the features. The cause of the lamination is

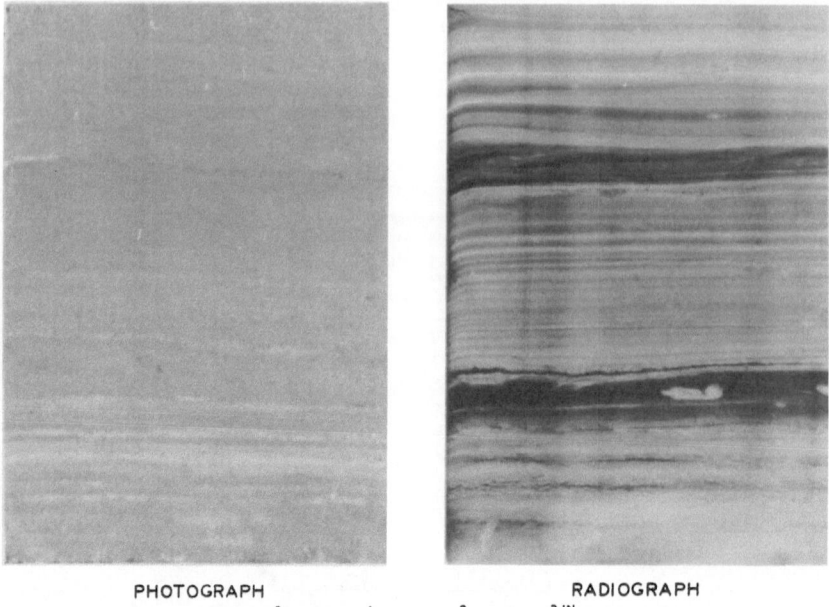

PHOTOGRAPH   RADIOGRAPH

0   1   2   3 IN.

Fig. 5-4. Varved clays in an interdistributary deposit along the Mississippi River near Head of Passes; depth 64 ft.

believed to be a combination of periodic variations in the influx of detrital materials and seasonal changes in conditions that favor the precipitation of carbonates. The clays are calcareous (the sharp white lines of the radiograph are believed to be local increases in the precipitation of carbonates). There may also be corresponding changes in the fabric of the clays.

Figure 5-4 shows another set of varved clays, though from an interdistributary deposit (see Table 5-1). Features described as determinative of subenvironments in the Atchafalaya backswamp deposits may, and do, exist in other deposits as well. Their use presupposes a sound understanding of sedimentation processes in all of the environments, and radiographic data helps to achieve such an understanding. The type of varving shown in Figs. 5-3 and 5-4 is rarely continuous for more than a few feet in vertical section and is discontinuous laterally; however, where it is continuous between borings, it furnishes superb correlation horizons.

Turbulence during deposition of clays may produce layers with highly contorted patterns sometimes called "convolute lamination." Figure 5-5 shows an example of such lamination. Note in the radiograph that such distortion lies between layers above and below that have been deposited under planar conditions. Similar features have been related to turbidity currents,

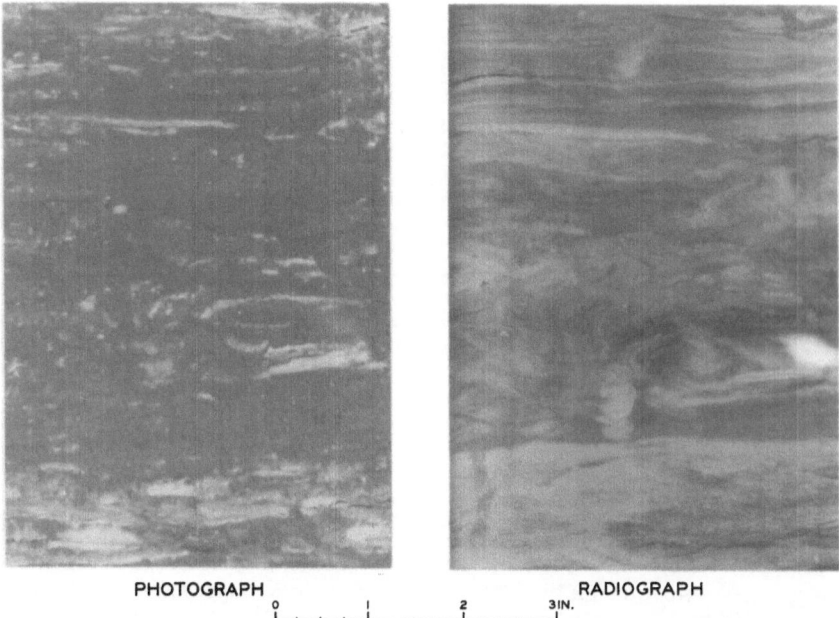

PHOTOGRAPH   RADIOGRAPH

0   1   2   3IN.

Fig. 5-5. Convolute laminations sandwiched between parallel layers, West Atchafalaya Levee Boring 31, Sample 26C; depth 107 ft.

abrupt changes in flow velocities, and intermittent storm or flood effects. No doubt they are the result of intense currents that are active temporarily.

In contrast, lake environments may also have long still periods of deposition. Clays deposited under such conditions can be entirely featureless, and clays of this type are found in the Atchafalaya backswamp. However, Fig. 5-6 shows such a clay from another environment—that of an abandoned channel along the Mississippi River. The radiograph confirms a complete absence of detail.

Stratified fine to medium sands occur in limited portions of the backswamp. Bedding may be parallel or cross laminated and truncated as shown in Fig. 5-7 from a silted channel deposit along the Mississippi River. Note that the radiograph shows finer details of stratification and truncation than does the photograph, and also shows secondary nodular concretions that are not evident in the photograph. It can be seen, however, that radiography is not as revealing of unobserved internal features in coarse detrital materials as it is in clays, although it is of considerable help in working out the details of sedimentation patterns. Figure 5-8 shows intricate changes in the laminar stratification and changes in the composition of detrital matter in a sand from a silted channel deposit. Such detail is not available in the photograph.

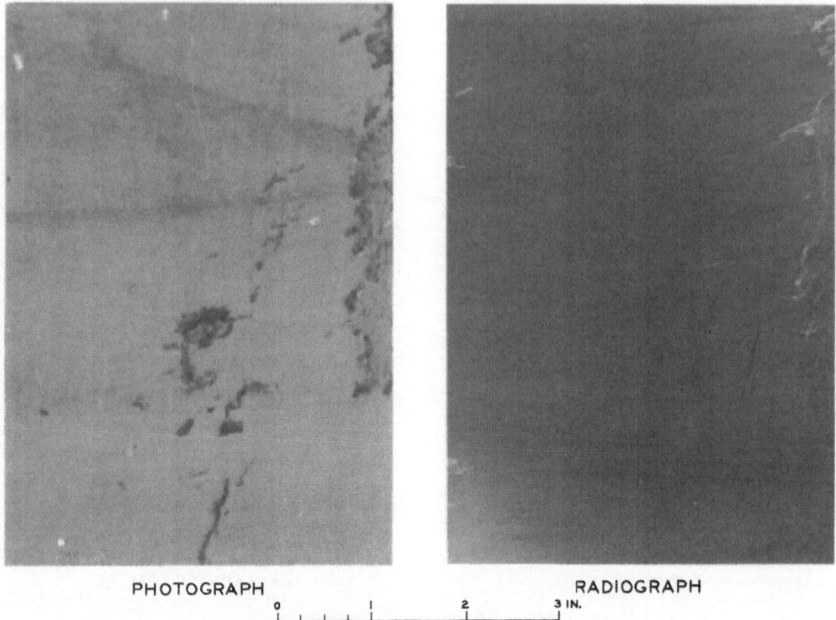

PHOTOGRAPH   RADIOGRAPH

0   1   2   3 IN.

Fig. 5-6. Featureless, massive clay deposited in still conditions of an abandoned channel deposit along the Mississippi River, Site 10; depth 79 ft.

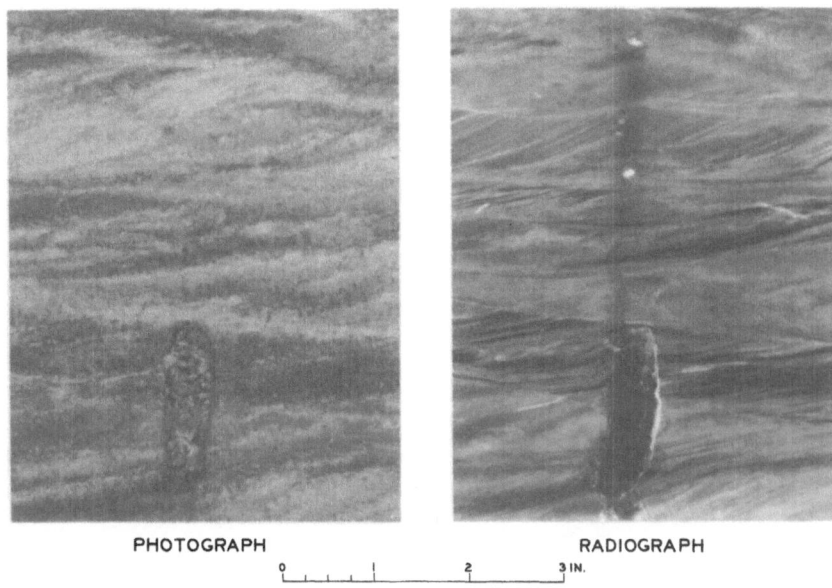

PHOTOGRAPH   RADIOGRAPH

0   1   2   3 IN.

Fig. 5-7. Complex stratification patterns in silty sand from silted channel deposit along the Mississippi River, Site 15; depth 172 ft.

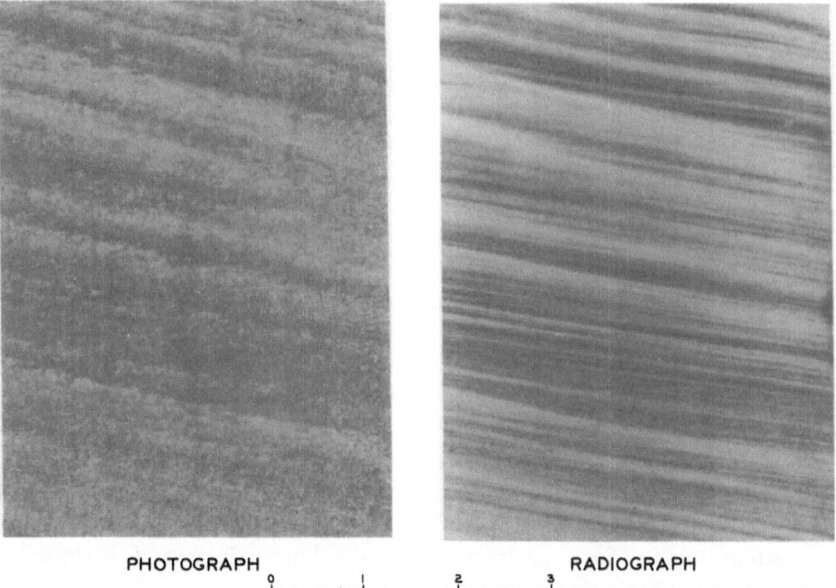

PHOTOGRAPH   RADIOGRAPH

0   1   2   3

Fig. 5-8. Details of variation in laminar stratification in silty sands in a silted channel deposit along the Mississippi River, Site 15; depth 144 ft.

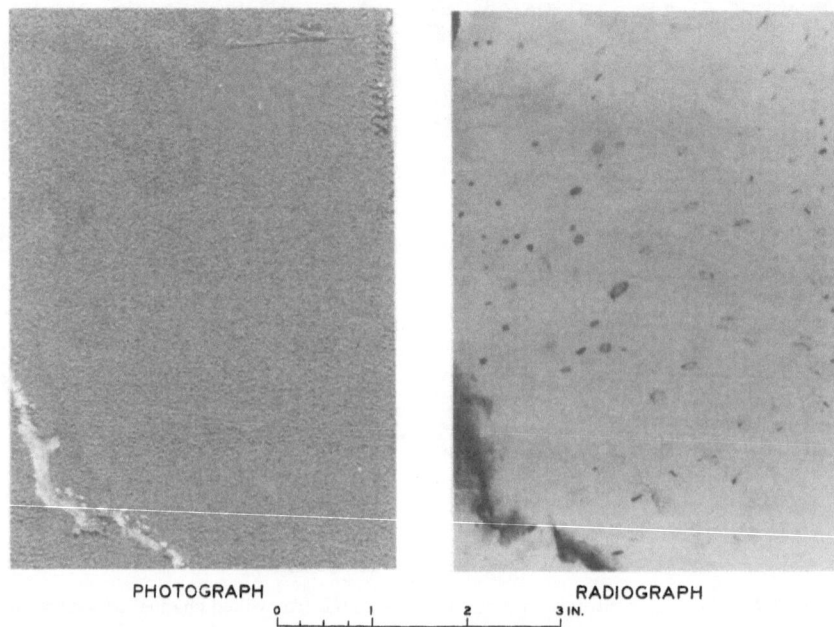

PHOTOGRAPH   RADIOGRAPH

0   1   2   3 IN.

Fig. 5-9. Massive sand with secondary ferruginous concretionary matter from a silted
channel deposit along the Mississippi River, Site 13; depth 91 ft.

Figure 5-9 shows the radiographic appearance of a featureless, massive sand,
also from a silted channel deposit but in this case confirming the massive
character and revealing secondary ferruginous concretionary matter not
seen in the photograph.

Where coarse deposits occur in the Atchafalaya backswamp, they are
believed to be relatable to deltaic facies formed by the entrance of streams into
the lakes. Other deposits of coarse-grained material may represent areas of
winnowing action in shallow lake-bottom flats that are not far removed from
the points of entrance of detritus carrying streams.

Invertebrate Fossil Remains. Invertebrate fossils in the backswamp are
mostly shells of bivalves and gastropods, the bivalves most commonly being
species of *Unio* and *Rangia*. Sometimes shells are so abundant that they form
"reefs" to thicknesses of several feet and are hundreds of feet in lateral extent. They may also be fragmental or occur as thin layers or lenses of shell
hash.

An example of fossil gastropods contained in the lake muds is shown in
Fig. 5-10. Sometimes the shells are rotted enough to be cut by the piano wire
but at other times the wire may pluck or dislodge the shells as is the case in
this figure. Figure 5-11 shows a shell hash from a nearshore gulf environment

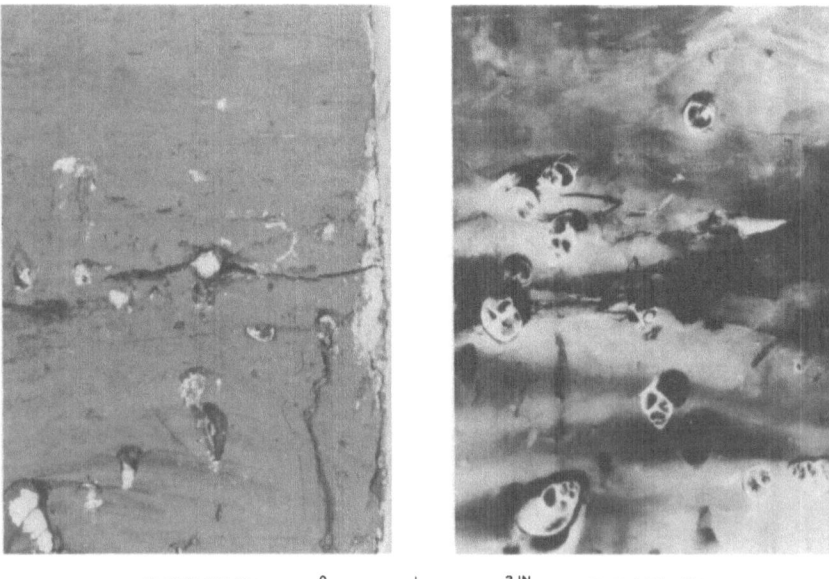

PHOTOGRAPH   0   1   2 IN.   RADIOGRAPH

Fig. 5-10.   Lake deposit containing fossil gastropods, East Atchafalaya Levee Boring 199, Sample 14c; depth 54 ft.

interspersed irregularly with clay; the radiography permits some of the detail to show which is absent in the photograph.

A common feature of lake deposits are burrows where invertebrates have churned through great thicknesses of clays; layers as much as 15 ft thick may be completely churned by this activity. Such burrowing introduces sands and silts into adjoining clay horizons. The appearance of burrows in a radiograph is shown in the upper part of Fig. 5-3.

Concretions.   Lake deposits are generally moderately calcareous and show an increase in carbonates with depth. Secondary carbonate concretions occur frequently and sometimes are abundantly developed in certain layers, existing in all stages of development from incipient mineralizations easily cut by piano wire to hard nodules several inches in length. The white area A in Fig. 5-3 is an incipient concretion that was cut by piano wire; another concretion shows up as a white area in the convolute laminations of Fig. 5-5.

Chemical analysis indicates that these concretions are combinations of calcite ($CaCo_3$) and siderite ($FeCO_3$) with siderite predominating. Coleman and Ho (1967) report that the amount of cementing materials increases with depth in the Atchafalaya sediments. These changes appear to accompany the consolidation process and may contribute to the growth of concretionary materials.

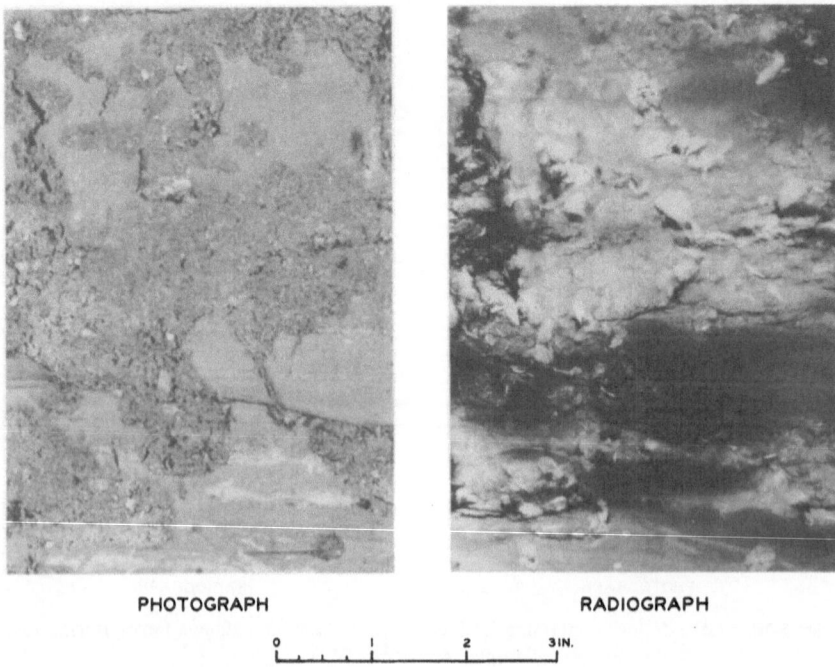

PHOTOGRAPH   RADIOGRAPH

0   1   2   3IN.

Fig. 5-11. Shell hash interspersed with clay in a nearshore gulf environment near Head of Passes, Mississippi River; depth 211 ft.

Organic Matter. Aside from organic oozes that are incorporated into clays, lake deposits may contain thin layers of organic matter interbedded with other stratified layers as shown in Fig. 5-12. Note the increased sharpness of bedded organic matter as shown in the radiograph. Such organic matter is reworked by wave action on the periphery of a lake where swamp materials are destroyed. In a boring, these reworked organics usually mark a transition zone between lake and swamp deposits. Also, stratified lake deposits may have penetrations from the longer roots of swamp deposits developed later from higher levels as shown in Fig. 5-13, where, in the radiograph, it may be noted that some of the roots have been mineralized with pyrite, the white-appearing roots, while others are nonmineralized. In both cases, the roots cannot be seen in the photograph.

WELL-DRAINED SWAMP ENVIRONMENT

Well-drained swamp deposits generally are characterized by the following features:

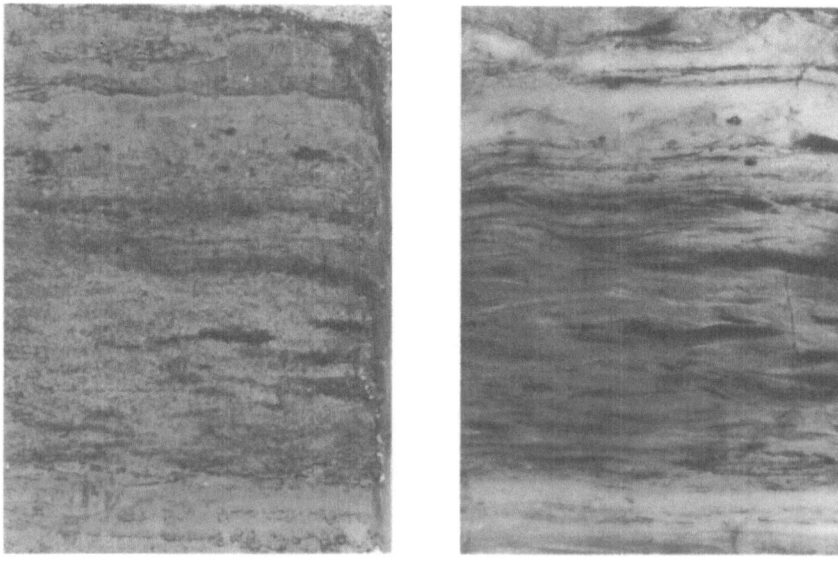

PHOTOGRAPH   0   1   2 IN.   RADIOGRAPH

Fig. 5-12. Bedded transition between lake and swamp deposits (note bedded organic matter), West Atchafalaya Levee Boring 29, Sample 25c; depth 98 ft.

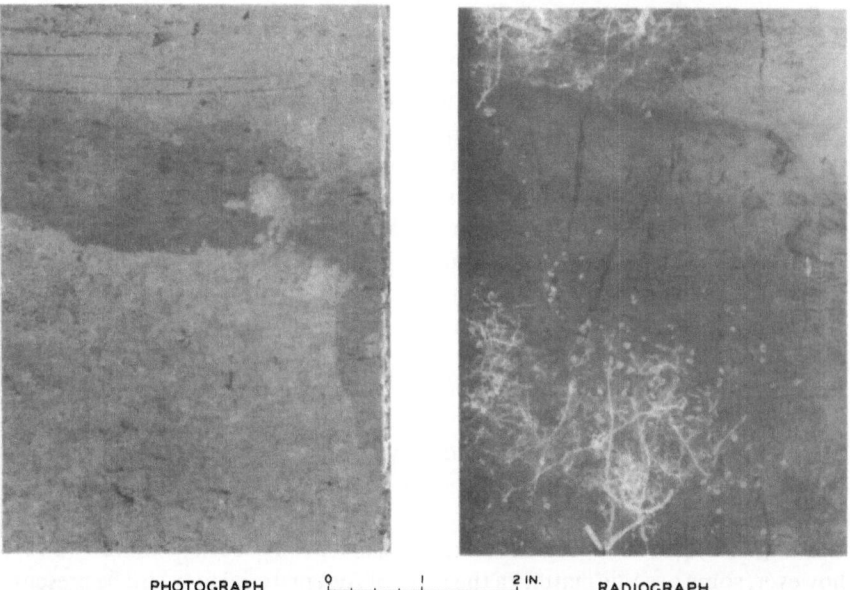

PHOTOGRAPH   0   1   2 IN.   RADIOGRAPH

Fig. 5-13. Transition between lake and swamp deposits, West Atchafalaya Levee Boring 39, Sample 10c; depth 38 ft.

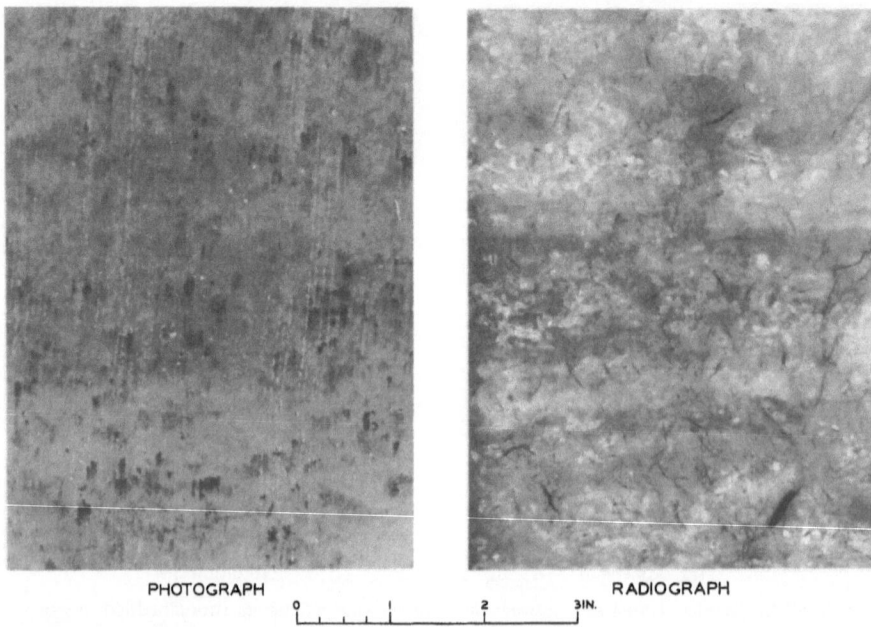

PHOTOGRAPH   RADIOGRAPH

0   1   2   3IN.

Fig. 5-14. Well-drained swamp deposit, East Atchalfalaya Boring 128a, Sample 2c; depth
7 ft.

Stratification.   Stratification is absent or is vaguely discernible as a result
of disturbance by a multitude of plant roots. Figure 5-14 shows a typical condition in which no bedding is seen in the photograph and is seen only vaguely
in the radiograph. The radiograph shows the full intricacy of root patterns,
but the organic matter of the roots has been largely removed by oxidation.

Color.   Color is light gray to buff to light yellow-brown and dark brown
depending on degree of oxidation and concentration of iron salts. Mottling
effects of iron oxides are common and, to a lesser extent, mottling from
manganese salts. The radiograph in Fig. 5-14 shows the intimate relation of
secondary mottling to the root patterns.

Concretions.   Concretionary matter is most commonly in the form of
iron oxide nodules and as iron oxide crusts around former roots. Black to
dark gray specks and amorphous concentrations of hydrous manganese oxides
are common.

Organic Matter.   Organic matter has usually been oxidized away,
however, some organic matter in the form of root material may still be present.
The presence of former roots is marked by fine root voids and iron oxide stains
and encrustments. Fissures representing blocky desiccation patterns form-

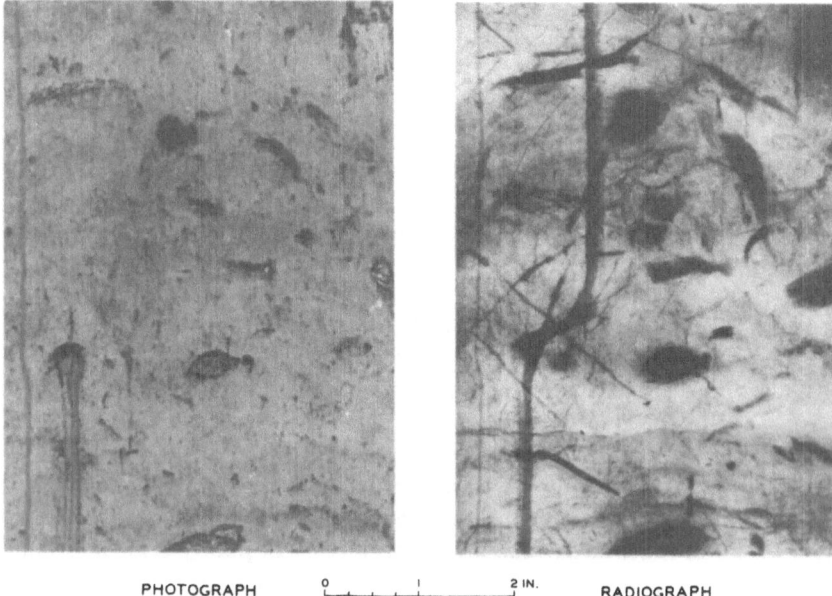

PHOTOGRAPH   0   1   2 IN.   RADIOGRAPH

Fig. 5-15. Poorly-drained swamp deposit of organic matter in clays, East Atchafalaya Levee Boring 128a, Sample 14d; depth 19 ft.

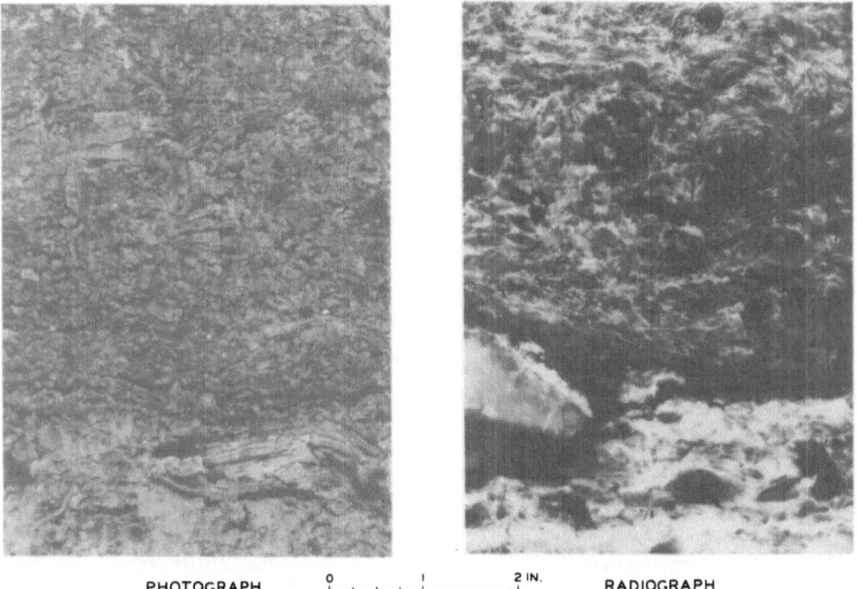

PHOTOGRAPH   0   1   2 IN.   RADIOGRAPH

Fig. 5-16. Peat in poorly-drained swamp, East Atchafalaya Levee Boring 201, Sample 7c; depth 26 ft.

ed from the drying of the soil by transpiration of plants are a common feature.

POORLY-DRAINED SWAMP ENVIRONMENT

Poorly-drained swamp deposits have the following characteristics:

Stratification. As with well-drained swamp deposits, stratification is either totally absent or is obscured by disturbance from root penetrations. Figure 5-15 shows the appearance of a typical poorly-drained swamp deposit. Note the extensive and intricate root penetrations in the clay matrix.

Color. Color is medium to dark gray to black where there is an excess of undecomposed organic ooze. Some organic matter may be brownish to black.

Organic Matter. Considerable organic matter may be preserved. Mostly it is in the form of remains of roots, however, bark, seed, pollen, twig, nut, and leaf imprints are common. Also common are organic oozes and detrital organic matter in lenticular masses. A special case of poorly-drained swamp is peat. Peat layers of nearly pure organic matter vary from inches to several feet in thickness; Fig. 5-16 shows the appearance of peat in a radiograph.

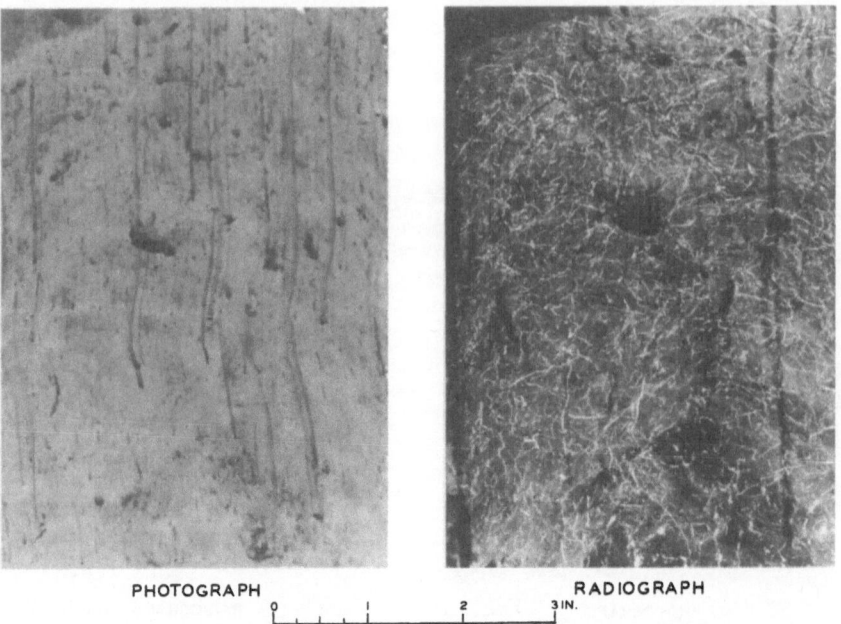

PHOTOGRAPH   RADIOGRAPH

0   1   2   3 IN.

Fig. 5-17. Poorly-drained swamp deposit; white lines in the radiograph are pyrite druses that have formed around roots, East Atchafalaya Boring 214, Sample 12d; depth 48 ft.

Radiography is not particularly helpful in the case of peat but the contacts with clay materials stand out more effectively than in the photograph.

Concretions. Concretions most commonly take the form of druses of pyrite (FeS_2) which form as small nodules in the clay and as encrustments around roots. The radiograph in Fig. 5-17 shows a striking example of pyrite druses formed around roots in the reducing environment of a poorly-drained swamp. Note that the structure is not visible at all in the photograph.

MICROSTRATIGRAPHIC CORRELATIONS

The preceding discussion showed the criteria by which radiographs were useful for subdividing sedimentary deposits in general, and in the Atchafalaya backswamp deposits in specific. The process may be termed microstratigraphy and the correlations it makes possible are microstratigraphic correlations. A further description of its workings may be had by referring to Figs. 5-18 and 5-19.

Figure 5-18 shows a selection of radiographs used in defining depositional environments in one boring, 214 UE, along the Atchafalaya Levee system. About twice as many radiographs were actually made for this boring. The interpretation in Fig. 5-19 shows three layers of lake environments, two layers of well-drained swamp, and one layer of poorly-drained swamp; also, the contact with levee fill is indicated. The boring log by itself provides a very insufficient basis for these separations and the engineering test data does very little to augment the lack of separable characteristics were the radiographs not available. However, with the radiographs, the separations are easy and convincing. Samples 7A and 7D show the broken-up configurations of a highly organic levee fill. Samples 10D and 11D lack stratification, are full of root patterns, and contain disseminated organic matter of the poorly-drained swamp. Samples 21D and 26D also have bedding obliterated by roots. In addition, they have oxidation nodules, and organic matter has been removed by oxidation. Thus, they mark well-drained swamp. The remaining samples show distinct bedding, varying layers of calcareous nodules, and other features of lake deposits.

Once the environmental horizons are established, it may be noticed that the test data are relatable to these horizons. Note that the water content increases in the swamp deposits and is less in the lake deposits. Within the swamp layers, the water content is greater in the poorly-drained swamp than in the well-drained. The trend in shear strength shows minor inflections that are relatable to the depositional horizons and so do the density values. However, it must be emphasized that these changes are subtle and they can be recognized and used only if borings have radiographic control. Fortunately such radiographic control can be projected to nearby borings where there

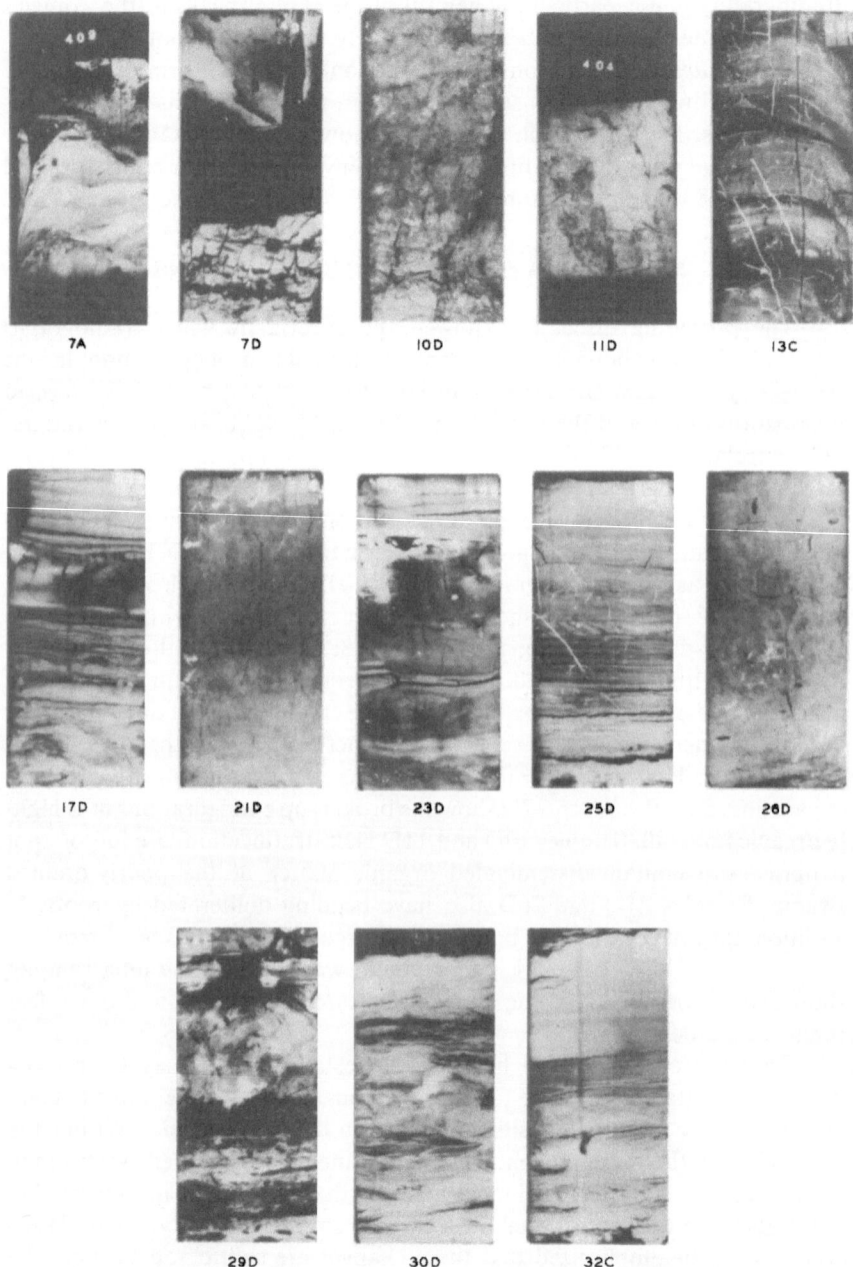

Fig. 5-18. Selected radiographs used in defining depositional environments in Boring 214 UE of the East Atchafalaya Levee. See Fig. 5-19 for interpretations.

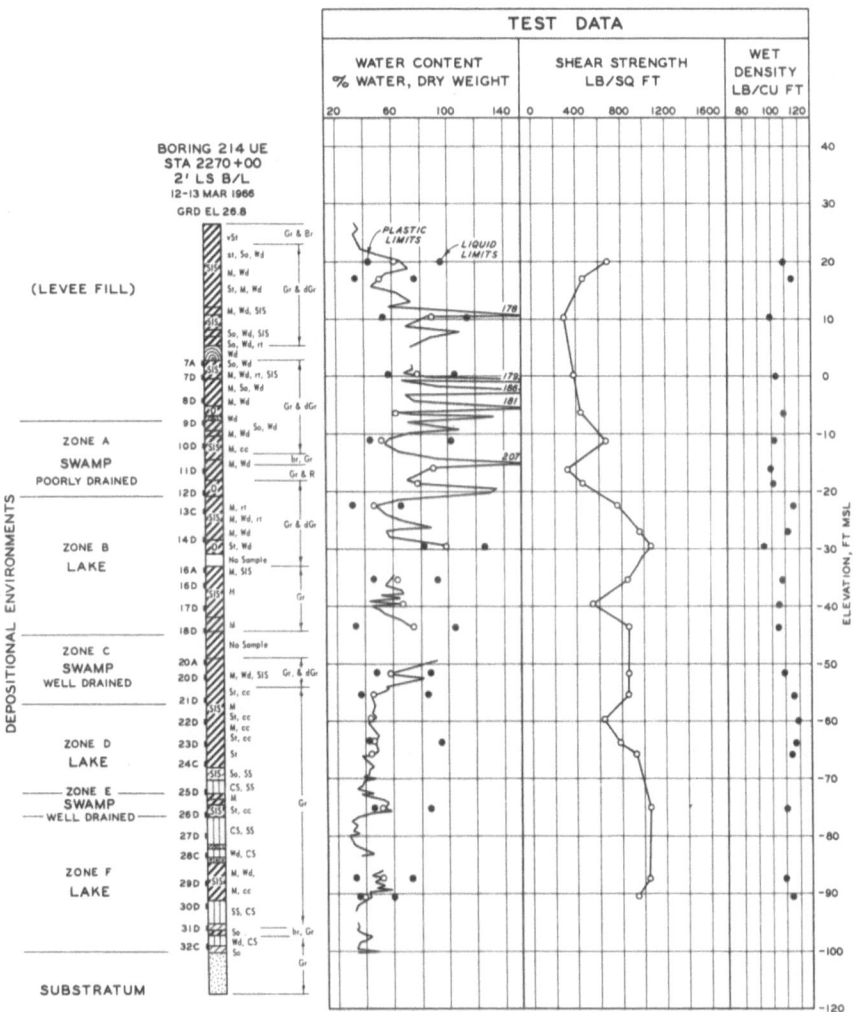

Fig. 5-19. Depositional environments and engineering test data for Boring 214 UE of the East Atchafalaya Levee.

are no longer any core samples available. Figure 5-20 shows an environmental interpretation based on the criteria developed by radiography but for an older boring where there are no longer any core samples on hand. Note that the separations between swamp and lake are sharply reflected in the water content, shear strength, and density.

Using the above method of combining borings with samples and projections into borings where there are no samples, correlations were made

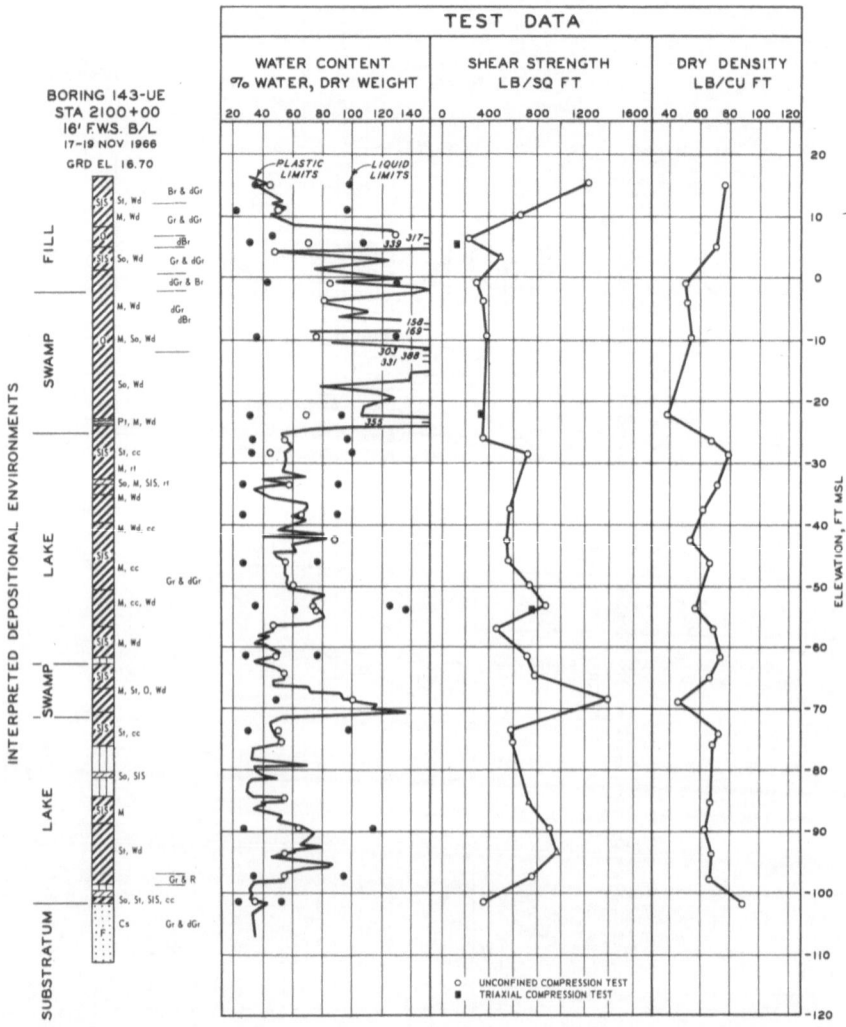

Fig. 5-20. Environmental interpretations based on projection, Boring 143 UE, East Atchafalaya Levee.

for most of the Atchafalaya Levee system. Figure 5-21 shows a segment of this correlation. Note how the level of settlement of the original ground surface has been traced where there is radiographic control and how the backswamp has been effectively subdivided into correlatable horizons. Such correlations are a first step in determining how the foundation soils contribute to settlement of the levees and what horizons may be more susceptible to settlement than others.

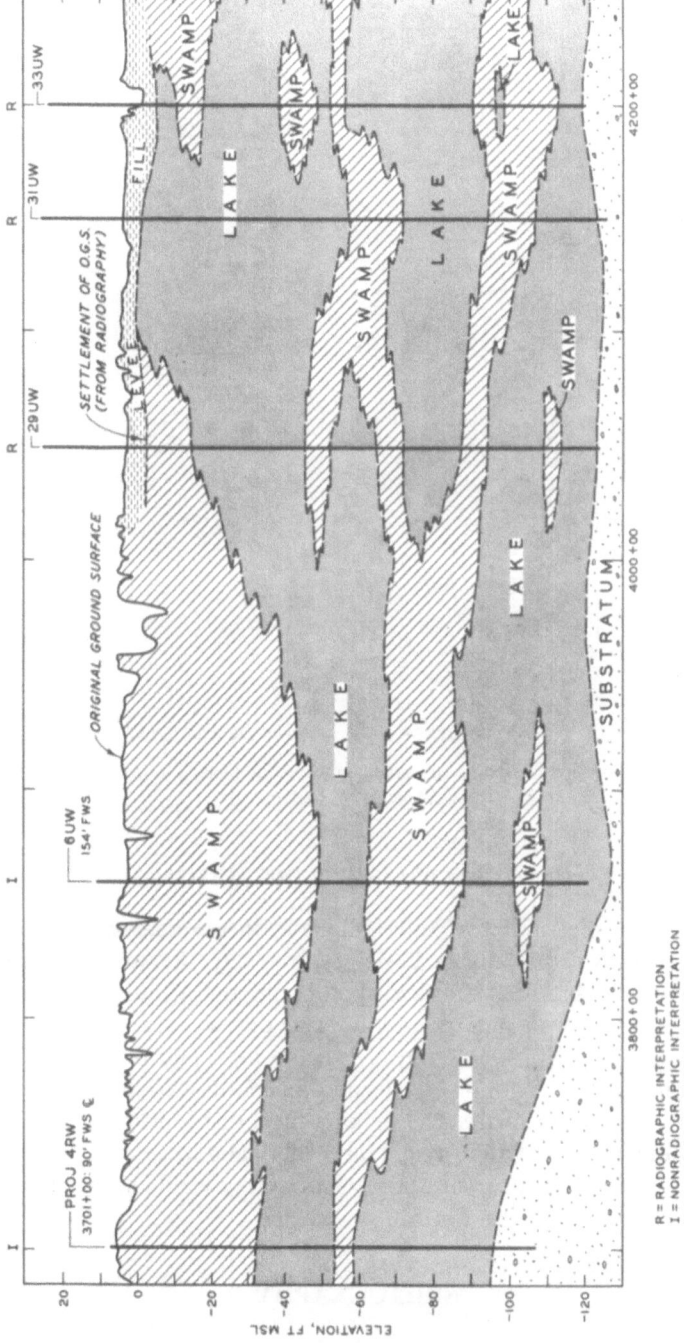

Fig. 5-21. Correlation along a segment of the Atchafalaya Levee system. Note control by radiography and projection through interpretation.

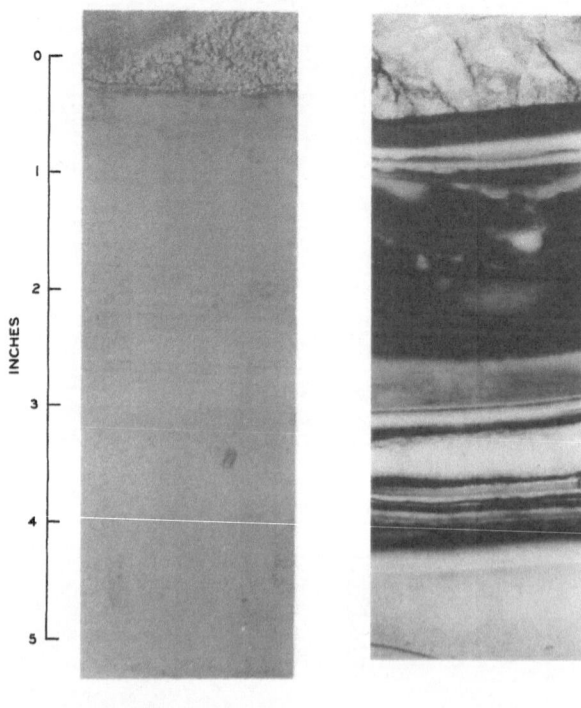

PHOTOGRAPH   RADIOGRAPH

Fig. 5-22. Radiographic detail of artificially sedimented Na-montmorillonite clay. (Courtesy J. M. Coleman.)

ARTIFICIAL SEDIMENTATION

A special application of radiography is in the analysis of artificially sedimented soils, especially in basic research into the strength properties of sediments. Figure 5-22 shows a photograph and a radiograph, made by J. M. Coleman, of a Na-montmorillonite clay that was sedimented artificially with changes in pH, introduction of calcium ions, and other changes. The absorption contrasts, with their reflection on density variations and strength variations, are extraordinary. The striking effect that is achieved in the radiograph is not apparent at all to visual inspection, and where laboratory studies are made on resedimented soils, x-ray monitoring would seem to be an important requirement.

BIBLIOGRAPHY

Bouma, A. H. (1963). Facies model of salt marsh deposits, *Sedimentology* **2**:122–129.

—— (1964a). Turbidities, in: A. H. Bouma and A. Brouwer (eds), *Turbidities*, Elsevier, Amsterdam, pp. 247–256.

—— (1964b). Ancient and recent turbidities, *Geol. Mijnbouw* **43**:375–379.

—— (1964c). Notes on x-ray interpretation of marine sediments, *Marine Geology* **2**(4): 278–309.

—— (1968). Distribution of minor structures in Gulf of Mexico sediments, *Trans. Gulf Coast Assoc. Geol. Soc.* **18**: pp. 26–33.

—— and N. F. Marshall (1964). A method for obtaining and analyzing undisturbed oceanic sediment samples, *Marine Geology*, **2**:81–99.

—— and F. P. Shepard (1964). Large rectangular cores from submarine canyons and fan valleys, *Bull. AAPG* **48**(2):225–231.

Calvert, S. E. and J. J. Veevers (1962). Minor structures of unconsolidated marine sediments revealed by x-radiography, *Sedimentology* **1**:287–295.

Coleman, J. M. (1966). Sedimentological characteristics and ecological changes in a massive fresh-water clay sequence, *Trans. Gulf Coast Assoc. Geol. Soc.* **16**: 169–174.

—— and Clara Ho (1967). Consolidation and cementation in sediments of southeast Louisiana, abstract, *Proc. 1967 Ann. Meeting GSA New Orleans*, pp. 37–38.

Hamblin, W. K. (1962). X-ray radiography in the study of structures in homogeneous sediments, *J. Sed. Petr.* **32**(2):201–210.

Kolb, Charles R. and R. I. Kaufman (1967). Prodelta clays of southeast Louisiana, in: *Marine Geotechnique*, Univ. Ill. Press, Urbana, pp. 3–21.

Krinitzsky, E. L. (1966). *Notes on X-Radiography of Soils*, special paper, Waterways Experiment Station, Vicksburg, Miss., 13 pp.

—— (1967). X-radiography for engineering—geological research in clays and clay shales, abstract, *Proc.* **1967** *Ann. Meeting GSA New Orleans*, p. 124.

—— (1970) *Correlation of Backswamp Sediments, Atchafalaya Test Section VI, Atchafalaya Levee System, Louisiana*, Waterways Experiment Station, Vicksburg, Miss. (in press).

—— and F. L. Smith (1969). *Geology of Backswamp Deposits in the Atchafalaya Basin, Louisiana*, TR S-69-8 Waterways Experiment Station, Vicksburg, Miss. 58 pp, 22 plates,

Rioult, M. and R. Riby (1963). Examen radiographique de quelques minerais de fer de l'Ordovicien normand. Importance des rayons X en sédimentologie, *Bull. Soc. Géol. France* **7**(5):59–61.

Shepard, F.P., R.F. Dill, and Ulrich von Rad (1969). Physiographic and sedimentary processes of La Jolla submarine fan and fan-valley, California, *Bull. AAPG* **53**(2):390–420.

Chapter 6

STRUCTURAL INTERPRETATIONS

The ability of x rays to reveal inhomogeneities in sediments that are not visible by reflected light has been described in the preceding chapter. This ability applies equally well to a large group of structural disturbances.

Radiography of the following examples was done by the same methods described in Chapter 5.

SHEAR FRACTURES AND DEFORMATIONS

Figure 6-1 provides a striking example of the ability of x rays to show fractures in clay that cannot be seen by reflected light. In this example, the photograph was taken after the sample had been air dried for a week and there was an opportunity for structure to appear that might have been revealed by drying. However, there is no substitute for the radiograph. Note that the fissures seen in the radiograph involve displacement of strata, and that these displaced layers are overlain by layers that are not displaced. Thus, the displacements must have occurred while sedimentation was taking place. At present, the horizon is 92 ft below the surface.

A nearly vertical shear displacement is shown in the varved clay of Fig. 6-2. Note in the photograph that there is a curved vertical break in the sample; this break also shows up in the radiograph but it is indistinct and was produced by handling of the sample. In the radiograph, however, there is a shear plane that displaces the delicate varves, which does not show up in the photograph. The importance of prior knowledge of these delicate shear planes in shear strength testing is obvious.

An example of complex fracturing at great depth in a large mass of prodelta clay is shown in Fig. 6-3. Note that radiography is essential to identifying the disturbances that have taken place. Though deformation has taken place in a highly plastic, fat clay, the clay has behaved much like a brittle material. At the same locality, but deeper, the bedding is seen to lie almost vertical. (Fig. 6-4). Again radiography appears to provide the only satisfactory method for determining the nature and extent of distortion in the clays.

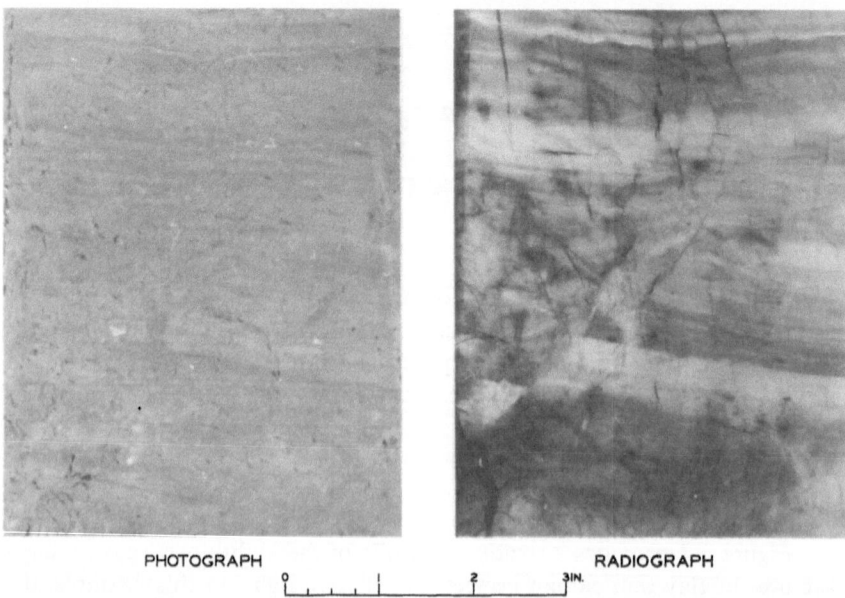

PHOTOGRAPH   RADIOGRAPH

Fig. 6-1. Fractures in lacustrine clays, East Atchafalaya Levee Boring 216, sample 23d; depth 92 ft. Fractures were developed contemporaneously with deposition.

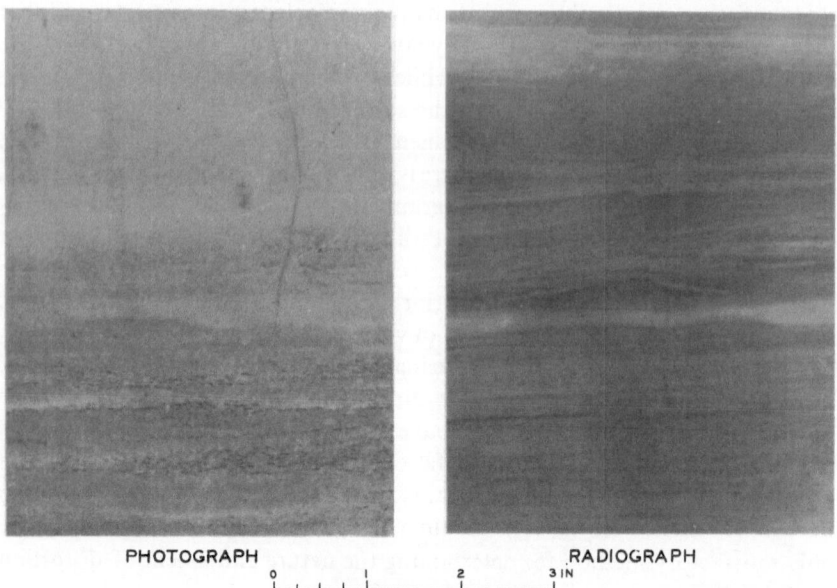

PHOTOGRAPH   RADIOGRAPH

Fig. 6-2. Vertical shear fracture in varved clay of channel fill deposit along the Mississippi River, Site 6; depth 47 ft.

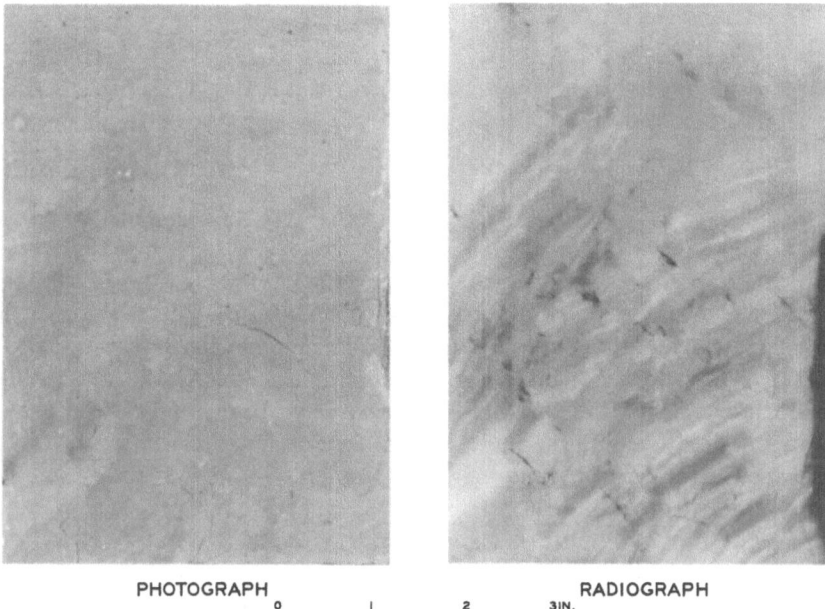

PHOTOGRAPH   RADIOGRAPH

0   1   2   3IN.

Fig. 6-3. Highly fractured and contented clays of the St. Bernard Prodelta, near Head of Passes along the Mississippi River; depth 118 ft.

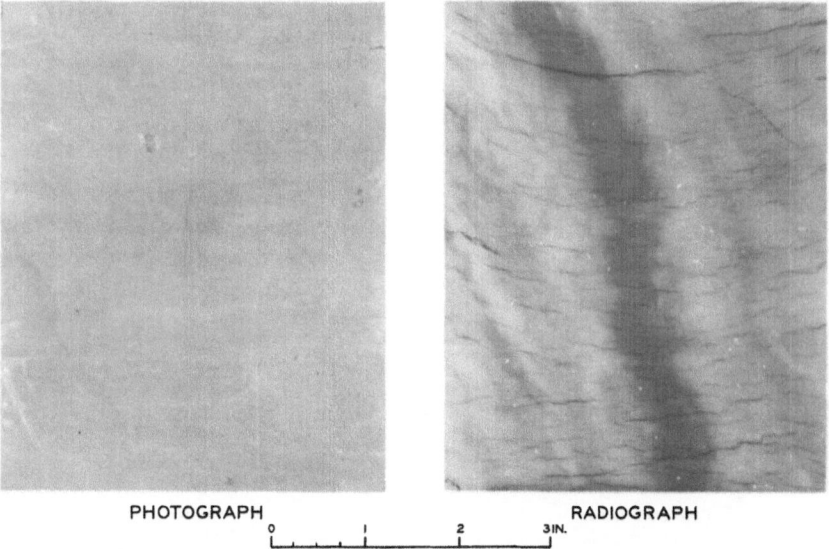

PHOTOGRAPH   RADIOGRAPH

0   1   2   3IN.

Fig. 6-4. Nearly vertical bedding planes in the pre-St. Bernard Prodelta, near Head of Passes along the Mississippi River; depth 198 ft.

EROSION SURFACES

Erosion by turbulence in a lake environment is shown in Fig. 6-5. The erosion surface is seen in the photography but the absence of the greater detail of the radiograph would make it hazardous to guess that it is strictly an erosion effect.

DESICCATION FRACTURES

The fissures in Fig. 6-6 are desiccation fractures that developed in lacustrine clays during exposure in a mudflat, though the horizon is now buried 43 ft below the surface. The radiograph is effective in illustrating desiccation effects, even though the cracks have been refilled with clay. Similar desiccation patterns are formed in some swamp deposits where roots remove water from clays through transpiration.

The above desiccation feature is datable as nearly contemporaneous with deposition. But, with burial, such features may occur deep in the section.

A special feature that results from desiccation, or some other process resulting in a loss of volume in the clay, is shown in Fig. 6-7. The short

PHOTOGRAPH   RADIOGRAPH

Fig. 6-5. Lake deposit disturbed by turbulence and containing root penetrations, East Atchafalaya Levee Boring 216, Sample 14c; depth 45 ft.

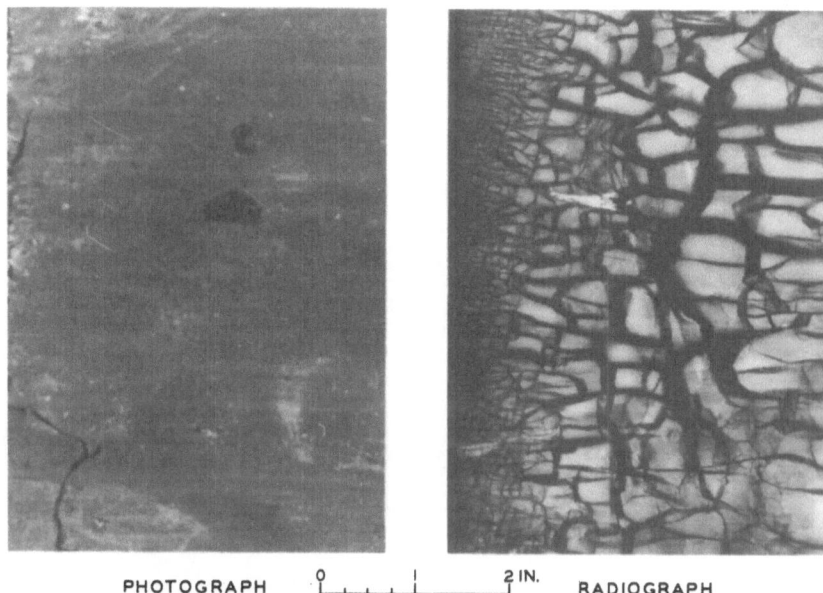

PHOTOGRAPH   0 ⌐————————⌐ 2 IN.   RADIOGRAPH

Fig. 6-6. Desiccation fractures in lacustrine clays, East Atchafalaya Levee Boring 205, Sample 11d; depth 43 ft.

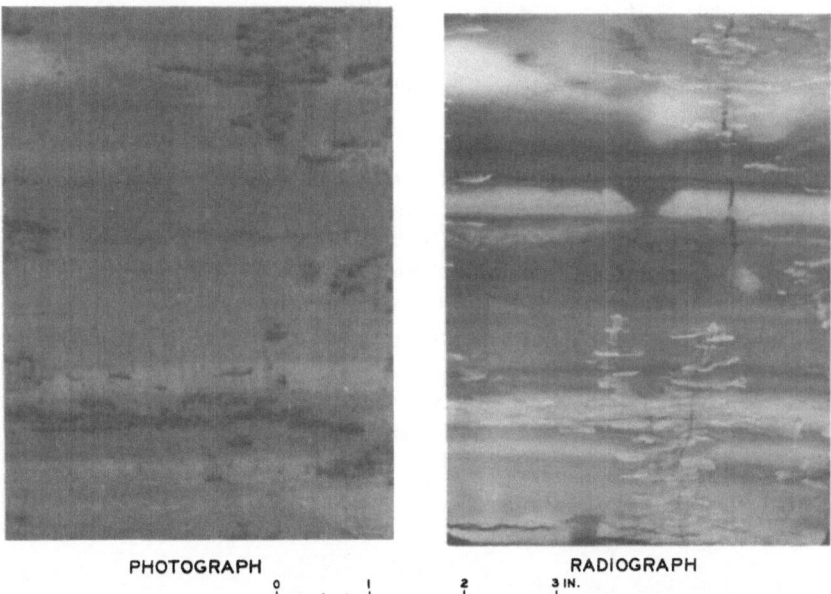

PHOTOGRAPH   RADIOGRAPH

0 ⌐————————⌐ 2   3 IN.

Fig. 6-7. Horizontal traces of disc fractures in clay, channel fill deposit along the Mississippi River, north of Vicksburg, Mississippi, Site 10; depth 70 ft.

Fig. 6-8. Example of disc fracture, channel fill deposit, along the Mississippi River, south of Mellwood, Arkansas, Site 8; depth 52 ft.

horizontal traces along the vertical fractures in the radiographs break open into disc shaped fractures, believed to be a form that develops to relieve tensile forces resulting from loss of volume. Figure 6-8 shows a photograph of a typical disc as it is commonly found in clays of the lower Mississippi Valley. In this case there is a rootlike opening near the center along which moisture was drained. Other samples may have stringers of coarse-grained sediment that facilitate drainage of the clays. Sometimes the discs develop prominently in certain layers of clays where there is no apparent connection with porosity or drainage. In undisturbed soil cores, these features are not likely to be recognized except with x rays.

VOIDS

Figure 6-9 shows open void spaces in a lacustrine clay where burrows by invertebrates have only partly collapsed. The layer is now 35 ft below the

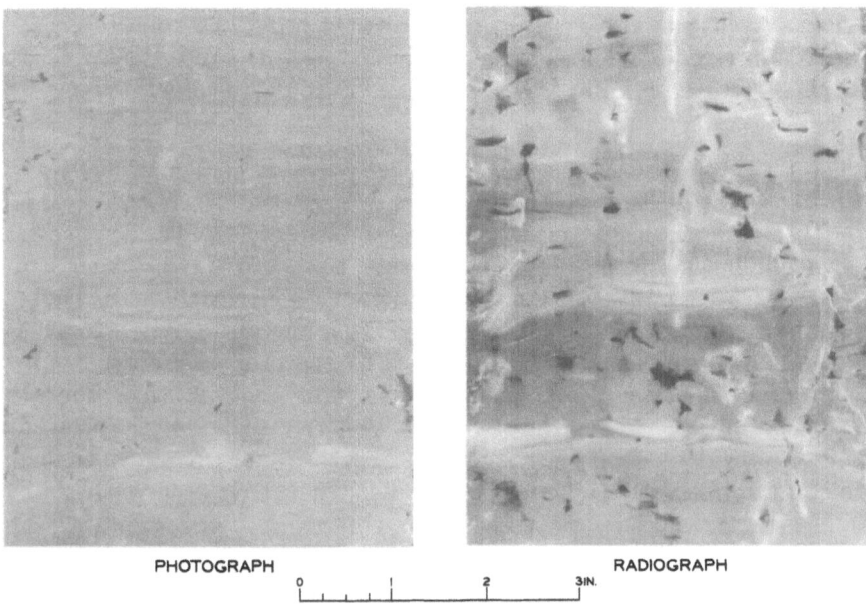

PHOTOGRAPH   RADIOGRAPH

0   1   2   3IN.

Fig. 6-9. Lake deposit containing open void spaces, East Atchafalaya Levee Boring 201, Sample 9d; depth 35 ft.

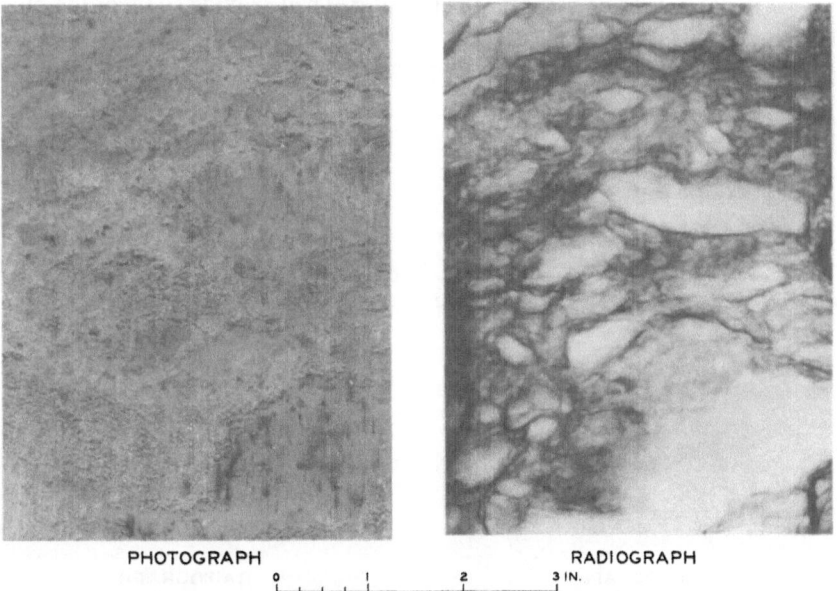

PHOTOGRAPH   RADIOGRAPH

0   1   2   3 IN.

Fig. 6-10. Remolded soil in levee fill along the Mississippi River at False River, Louisiana, Site 13; depth 16 ft.

surface. Voids may also be left by roots that have rotted or a void space may be occupied by decomposed organic matter having little or no strength. Such features are present in the backswamp deposits to depths of 40 ft or more.

REMOLDED SOILS

An important feature of radiography is its capacity to delineate remolded conditions in soils. Figure 6-10 shows the detail that may be found in a levee fill. The ease with which fill characteristics can be recognized by radiography has important uses in soils engineering interpretations. It permits accurate observations of contacts between fill and original ground surface and, where settlement of the original ground surface has occurred, provides an accurate measure of total displacement. Within levee fill or other reworked soils, it provides a supplementary means to evaluate the soil structure that has been formed and to judge if desired conditions have been established.

CLAY SHALES

Clay shales are defined for the purpose of this discussion as lithified sedimentary material consisting mostly of clay minerals that are subject to slaking action when saturated with water. They provide some of the most treacherous problems for stability of foundations and slopes. Part of the problem of dealing with clay shales is that visual examination of fresh cores

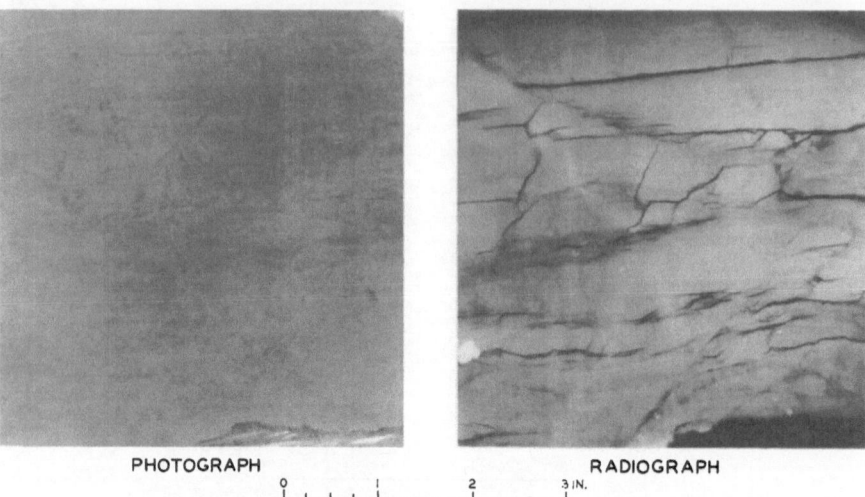

PHOTOGRAPH   RADIOGRAPH

0   1   2   3 IN.

Fig. 6-11.   Deformation in Pepper shale, Waco Dam site, Texas; depth 45 ft.

and subsequent laboratory testing do not sufficiently reveal their potential susceptibilities to failure. Some recent experiments show that radiography can provide some facets of information on past disturbances in clay shales that might not be available otherwise and that might reveal their susceptibility to further disturbance.

Compare the photograph and radiograph in Fig. 6-11. The sample is Pepper shale from the foundation of Waco Dam in Texas. Preparation for radiography was made by sawing off 5-mm-thick slices on a bandsaw. Exposure was the same as for the preceding samples. Waco Dam is the site of a slide in the Pepper shale that took place underneath the dam as the dam was being built (Beene, 1967). Part of the problem was that preconstruction exploration did not reveal the existence of zones of weakness in the underlying shale. Figure 6-11 helps to explain why zones of weakness or of progressive deformation were hard to find. Clay shales are plastic and though they may deform as brittle materials, they tend to seal the planes along which deformation takes place. The radiograph in Fig. 6-11 shows three intersecting sets of fracture planes and an undulation caused by deformation of horizontal bedding planes. None of this is seen in the photograph.

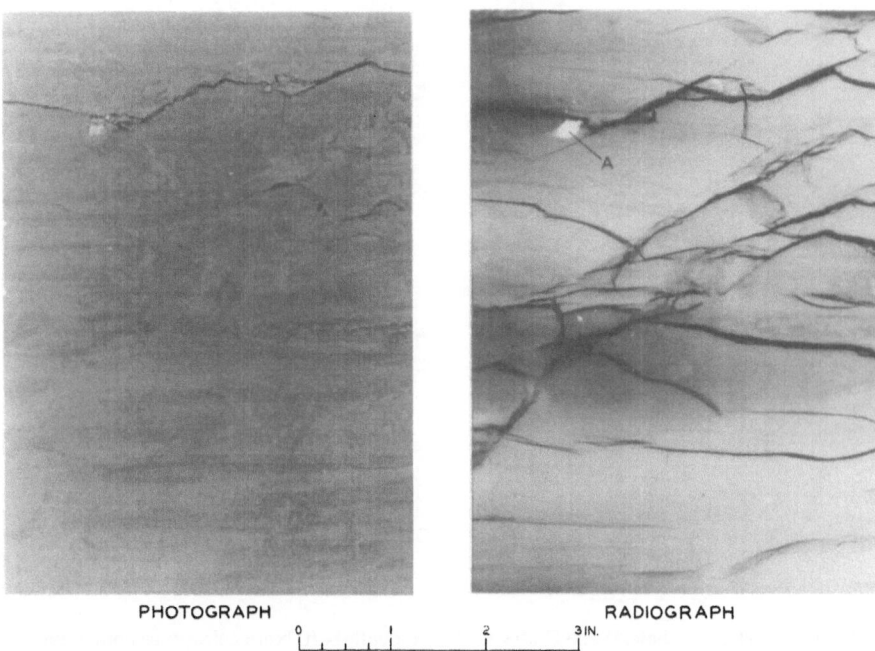

PHOTOGRAPH   RADIOGRAPH

Fig. 6-12. Shear zone in Pepper shale, Waco Dam site, Texas; depth 47 ft. A is siliceous concretion.

Figure 6-12 shows a further comparison between photograph and radiograph of a Pepper shale with a pronounced slippage zone. Part of the disturbance, in this case, can be seen in the photograph, but not enough to make out the full extent of the traces in the radiograph.

Detail in a portion of the Del Rio shale at Waco Dam site is contained in Fig. 6-13. Note the zones of calcareous concretionary matter with open solution cavities, the tendency to horizontal fissures along the bedding planes, and the absence of other fracture patterns. This sort of detail is important in investigating the properties of clay shales as directions and intensities of fracture zones and points of entrance of water are critical to design considerations. Also, assignment of laboratory tests needs to be related to the presence and absence of disturbances. Critical horizons should not be missed. Radiography may not have all of the solutions but it can provide an additional category of relevant information.

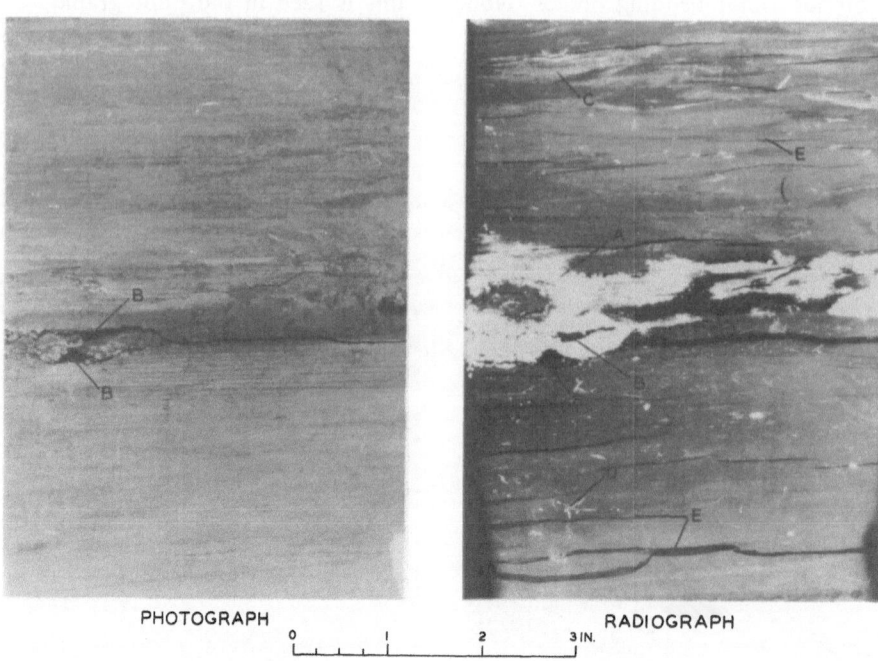

PHOTOGRAPH   RADIOGRAPH

Fig. 6-13. Del Rio shale, Waco Dam site, Texas; depth 14 ft. Note calcareous concretionary matter (A and C); open solution cavities (B); calcareous matter of organic origin (D); horizontal fissures along bedding planes (E).

BIBLIOGRAPHY

Beene, Ralph R. W. (1967). Waco Dam slide, *Proc. ASCE, Journal Soil Mechanics and Foundation Division* **93** (SM4): 35–44.

BIBLIOGRAPHY

Chapter 7

CORE SCANNING

Scanning of unopened or partly opened sediment cores by x radiography has been so widely applied recently that it is now a routine procedure in some operations, particularly in oceanography. However, publications describing the results of such work are few. Notable among extant papers are those by Stanley and Blanchard (1967), Stanley and Kelling (1967), and Bouma (1968) dealing with oceanographic cores. Haase (1967) described radiographic results from unopened soil cores taken in backswamp deposits of the Atchafalaya basin in Louisiana. Rukavina (1967) reported on radiography of split cores taken in Lake Ontario sediments. All of these reports confirm the practicality of using x radiography as a scanning and recording technique for the study of sediments.

X-ray scanning of cores taken in hard rock would not be nearly as useful. Hard-rock cores are not as perishable as are fine-grained sediments and they can be stored without being elaborately sealed. Their features are also more directly visible and x rays are not needed to provide added dimensions as is the case in unconsolidated or parly consolidated materials. Consequently, x-ray scanning of hard rocks is not usually productive of unique information.

An intermediate area in which radiography is useful is in the study of clay shales. So far there have been no published reports in this area, but some experimental work has been done at the Waterways Experiment Station on clay shales from Waco Dam, Texas (see Chapter 6), from the Missouri River Valley, and from the Panama Canal Zone. These observations suggest that radiography is helpful in delineating shear zones and in locating horizons with secondary mineralizations. Such information can be useful as an adjunct to testing programs in a soils laboratory.

Radiography of whole cores can be done without any special preparation. Stanley and Blanchard (1967), Stanley and Kelling (1967), and Rukavina (1967) have published such radiographs. Exposures can also be made of samples that are still in steel or plastic tubes or encased in paraffin, but better results are obtained if compensations are provided for the cylindrical shape of the cores. X rays tend to burn the film where they pass through the narrower chords of the cylinder and there is also a problem with scatter and undercutting from radiation along the sides. Compensation for cylindrical shape can be made by setting the cores in plaster-of-paris molds shaped to produce a square or rectangular cross section. Where molds are too cumbersome for use in the handling process, curved lead plates (see Fig. 7-1) can be used to reduce scatter and undercutting.

Bouma (see Fig. 7-2) developed a simple and inexpensive technique for radiographing cores that are split through the middle. The halves are laid in a trough where they are imbedded in sand as compensation for their half-cylinder shape. Groups of these cores are radiographed in four exposures on four sheets of 14×17 in. film. The setup can be operated on shipboard. Other workers have been reported to take stereo views of whole cores, notably at the University of Georgia Marine Institute at Sapelo Island, as a routine procedure.

A study was conducted at the Waterways Experiment Station (Haase, 1967) to determine to what degree of reliability interpretations could be made of radiographs taken of 5-in-diameter soil cores encased in paraffin in a cardboard cylinder. The total cylinder was $6\frac{1}{8}$ in. in diameter. Cores were from the Atchafalaya Basin in Louisiana. Radiographs were made of the unopened core, a $4\frac{1}{4}$-in. slice, a 3-in. slice, a 2-in. slice, and a $\frac{3}{8}$-in. slice, all taken from the same samples. Figure 7-3 illustrates comparative radiography from a poorly-drained swamp facies in the backswamp of the Atchafalaya Basin. Note

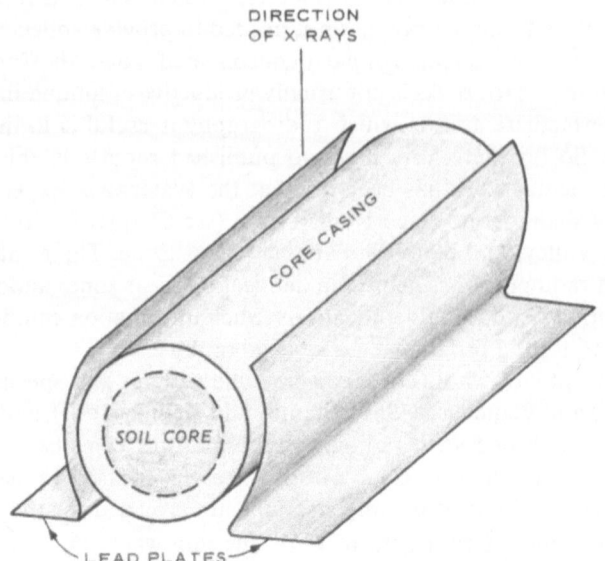

Fig. 7-1. Use of lead plates to reduce scatter and undercutting.

Fig. 7-2. Picker "Hotshot" x-ray unit with trough for imbedding split cores in sand. (Courtesy Arnold H. Bouma.)

that radiographic detail of large roots and of fine roots is best in the 2-in. slice. Both features are diminished in the ⅜-in. and thicker slices. The features are faint in the unopened core, However, registration is nonetheless sufficient that a worker with experience would have no difficulty in making a satisfactory interpretation. An intricate pattern of fine roots with iron oxide stains around the roots is shown in Fig. 7-4. The root pattern is recognizable in the unopened core but the secondary mineralization is not. Stratification and burrows were examined in the radiographs shown in Fig. 7-5. The stratification within the upper layer of sand shows up effectively, and the contact between

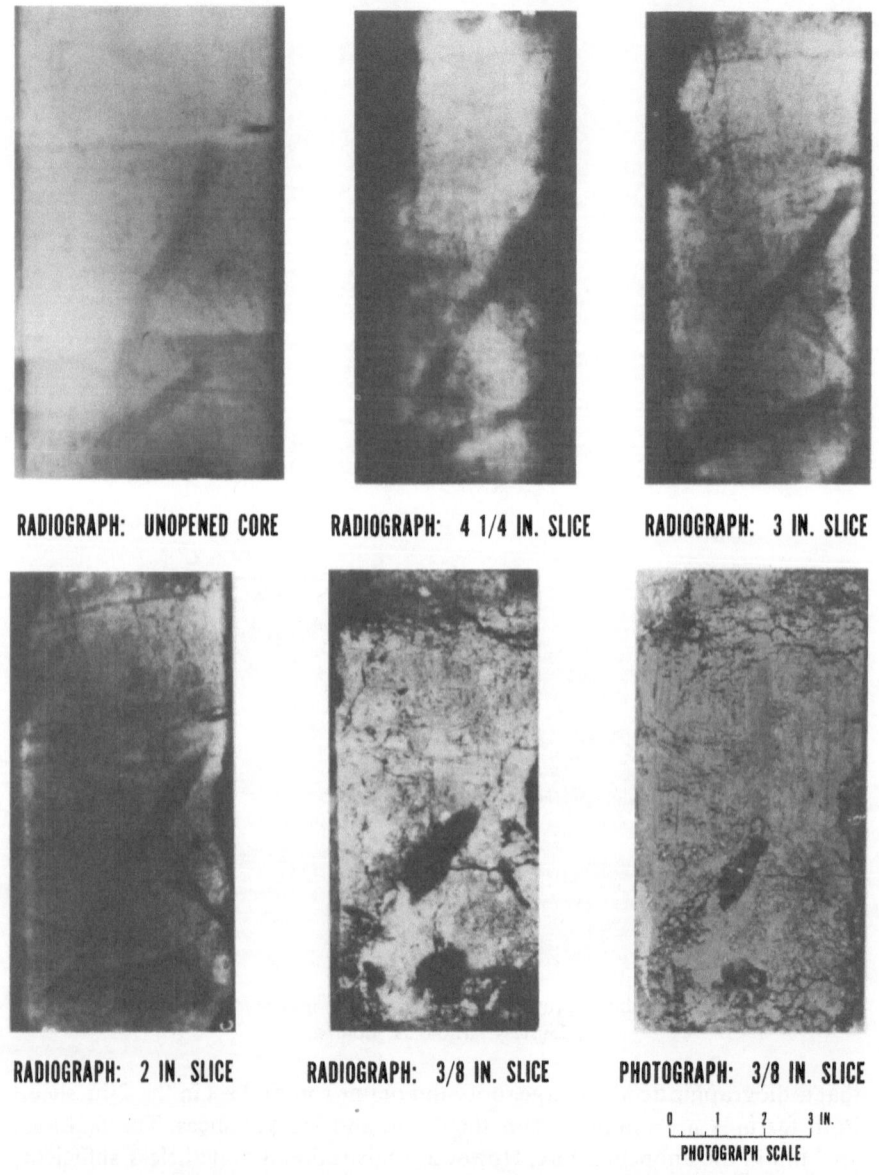

RADIOGRAPH: UNOPENED CORE   RADIOGRAPH: 4 1/4 IN. SLICE   RADIOGRAPH: 3 IN. SLICE

RADIOGRAPH: 2 IN. SLICE   RADIOGRAPH: 3/8 IN. SLICE   PHOTOGRAPH: 3/8 IN. SLICE

0   1   2   3 IN.
PHOTOGRAPH SCALE

Fig. 7-3. Comparative radiography of unopened core and slices containing large roots
and fine roots in backswamp deposits of the Atchafalaya Basin, La.

the sand and clay is distinct. The burrow in the clay also shows up, faintly but
sufficiently for interpretation. Figure 7-6 shows details of bedding and Fig.
7-7 shows slickensided fractures in unopened cores from the same area.

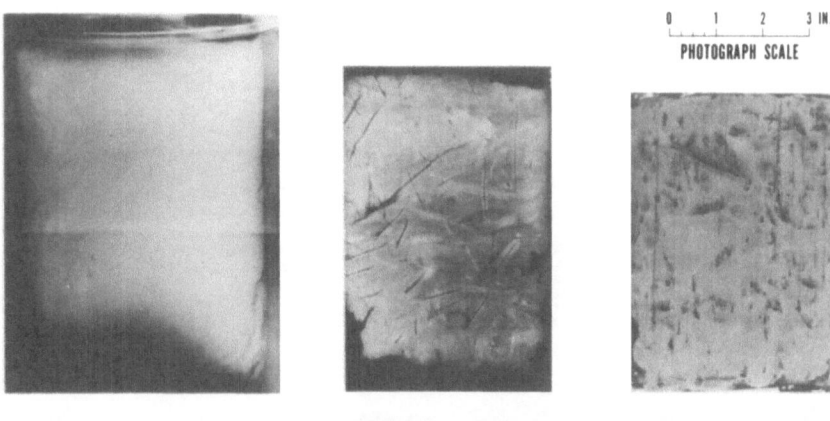

RADIOGRAPH: UNOPENED CORE   RADIOGRAPH: 3/8 IN. SLICE   PHOTOGRAPH: 3/8 IN. SLICE

Fig. 7-4. Radiography of a core containing a pattern of small roots with secondary iron oxide mineralization around the roots, Atchafalaya Basin, La.

The radiography was done with a Norelco 100 kV beryllium window tube using a focal distance of 33 in. and Kodak type M film. Exposures were at 95 kV and 10 mA. Variations were made in the times of exposure.

Haase was able to generalize that the type of sediment (clay or sand) and the presence of appreciable quantities of organic matter were easily

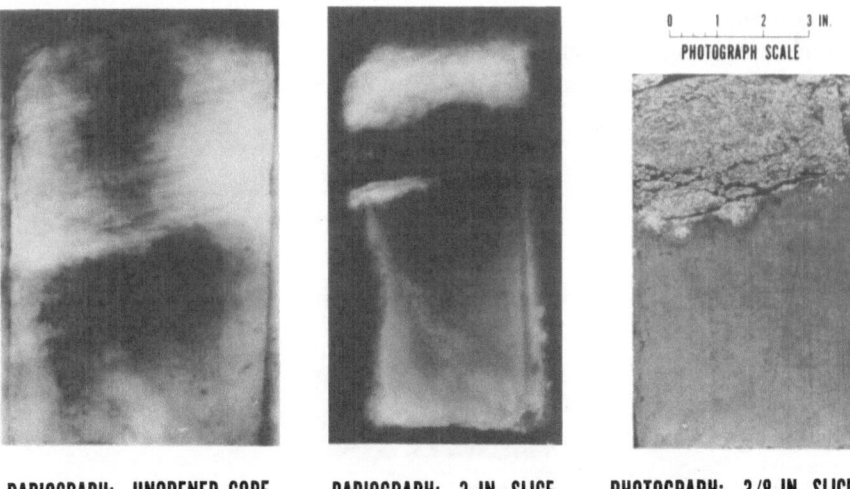

RADIOGRAPH: UNOPENED CORE   RADIOGRAPH: 3 IN. SLICE   PHOTOGRAPH: 3/8 IN. SLICE

Fig. 7-5. Radiography of a core showing stratification and burrows, Atchafalaya Basin, La.

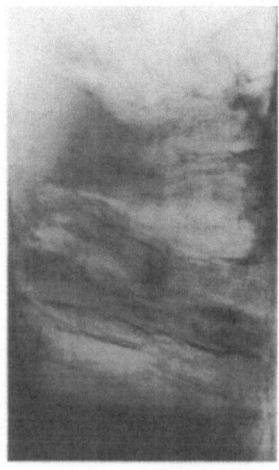

RADIOGRAPH: UNOPENED CORE

Fig. 7-6. Radiograph of a complex bedding in an unopened core from the Atchafalaya Basin, La.

recognized. Plastic deformation, fractures, and burrows filled with a contrasting material were recognizable though not as easily. Concretions or secondary mineralizations were identifiable only when they were large. Very large roots and some fractures could be seen better on radiographs of unopened cores than on thin slices destroyed in the cutting process. Some depositional features, such as turbulence in the sedimenting of clays and some burrows, were not determinable.

In considering core samples, radiography of other bulk samples should

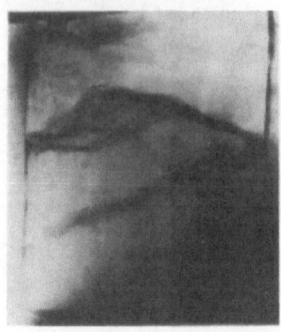

RADIOGRAPH: UNOPENED CORE

Fig. 7-7. Radiograph of slickensided fractures from an unopened core from the Atchafalaya Basin, La.

be mentioned. These include box samples taken either in test pits or on the ocean floor. Quite often a slice can be taken from a large block and radiography may then provide information that is useful in interpreting the larger

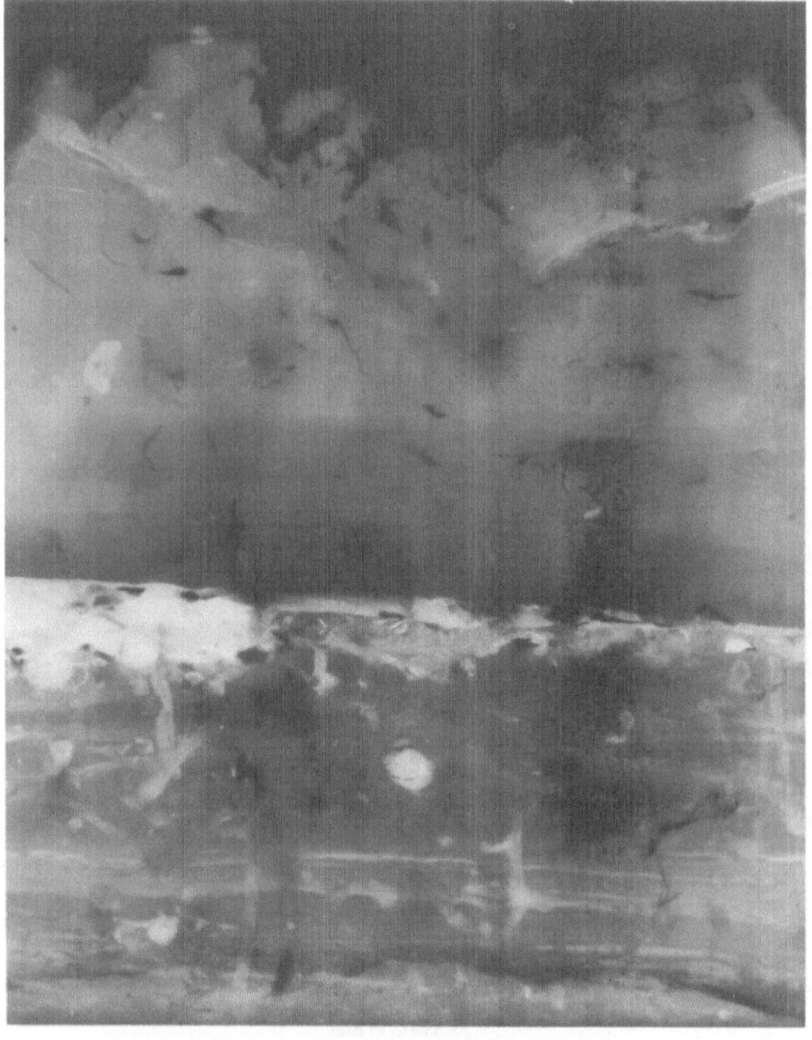

5 INCHES

Fig. 7-8. Radiograph of an erosional contact between soft clay, above, and stiff clay, below. From the ocean floor off Surinam. (Courtesy A. H. Bouma.)

5 INCHES

Fig. 7-9. Radiograph of clay with turbulent bedding patterns overlying a sandy gravel.
From San José Canyon, Baja California. (Courtesy A. H. Bouma.)

specimen and selecting the procedures for further tests. Two exceptionally fine radiographs made from box samples of ocean floor materials are shown in Figs. 7-8 and 7-9. These were prepared by Arnold H. Bouma. Figure 7-8 shows an erosion contact which has formed on a stiff clay with horizontal laminae of sandy silt material. The clay is considerably marked by burrows. Numerous small invertebrate shells occur within a few inches of the erosion surface, which is overlain by a very soft clay containing some fine burrows and a few shells. The box sample was taken off the coast of Surinam.

Figure 7-9, made from a box sample taken in the San José Canyon in

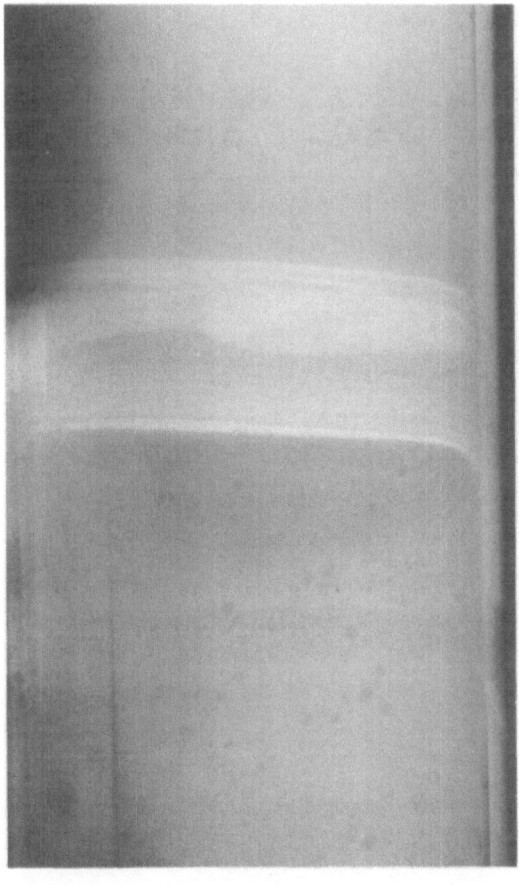

1 INCH

Fig. 7-10. Turbidite with ripple marks. Gulf of Mexico. Detail from split-core radiography. (Courtesy A. H. Bouma.)

Baja California, shows a sandy gravel overlain by a clay which is marked considerably by turbulence during deposition. These radiographs were made from slices 1 cm thick. Exposures were at 40 kV and 5 mA for 3 min at a focal distance of 38 in. The Surinam sample was radiographed on Cristallex film from Kodak, Ltd. (England). The San José Canyon sample was radiographed on Kodak type AA film.

Figures 7-10 to 7-14 show radiographic details selected from routine scanning of split cores taken in cruises of the R/V Alaminos in the Gulf of Mexico. Figure 7-10 shows a turbidite layer contained in material laid down

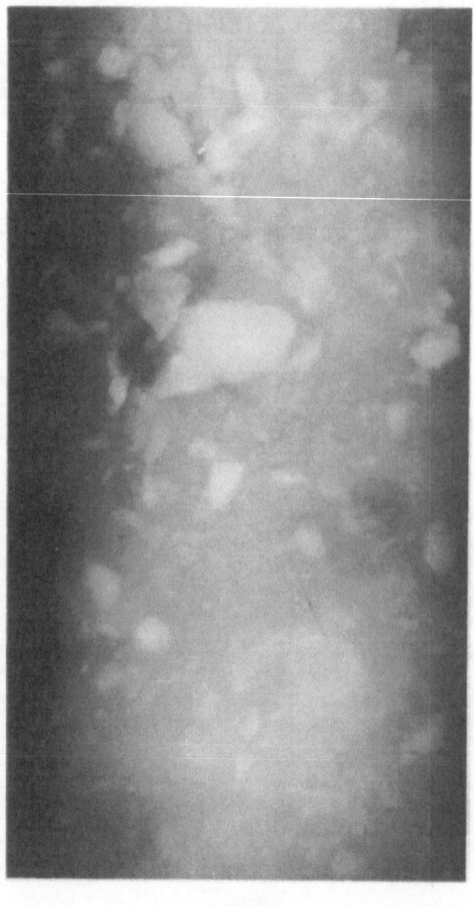

I INCH

Fig. 7-11. Breccia from submarine slumping. Gulf of Mexico. Detail from split-core radiography. (Courtesy A. H. Bouma.)

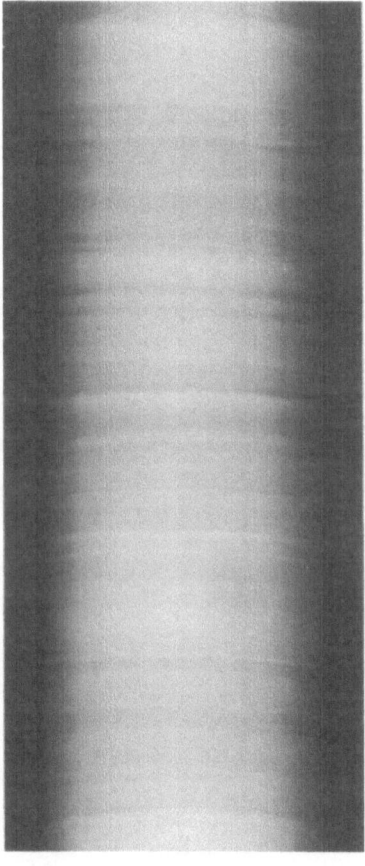

I INCH

Fig. 7-12. Varve-like laminations. Gulf of Mexico. Detail from split-core radiography.
(Courtesy A. H. Bouma.)

in still, continuous deposition. The turbidite layer has both planar bedding
and well-developed ripple marks. Figure 7-11 shows brecciated material possibly developed in a zone of submarine slumping. Figure 7-12 shows horizontal laminations of varve-like fineness with a cyclical pattern of sequential
deposition. Figure 7-13 shows irregular mycelia distributed through most of
the sample. Figure 7-14 shows ruptures which are believed to have been formed by expansion of H_2S gas, which occurred with the decrease in pressure when
the sample was brought to the surface.

Fracture patterns imposed by expanding gases, as above, or by piston

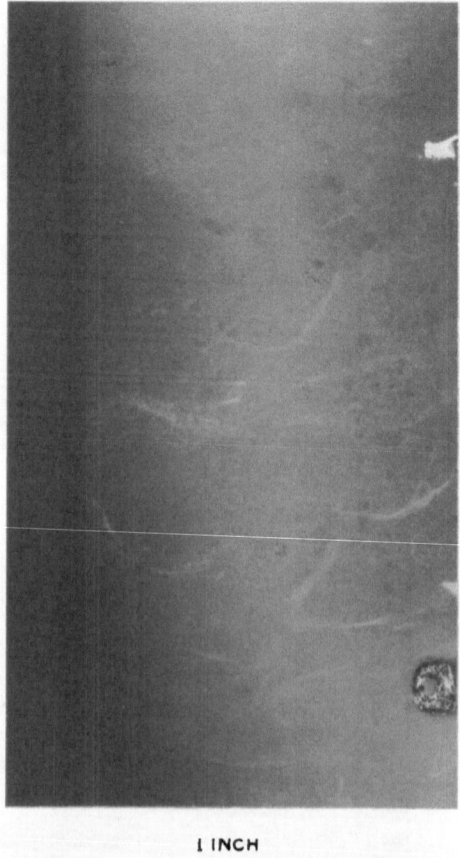

I INCH

Fig. 7-13. Mycelia. Gulf of Mexico. Detail from split-core radiography. (Courtesy A. H. Bouma.)

extrusion of cores from their cylinders, are usually easily recognized with a moderate amount of experience. Piston extrusion of some materials may induce parabolic patterns of bending and short, discontinuous fracturing. Usually, such effects form symmetrical patterns around a centerline and are recognized by their symmetry.

Bouma and Boerma (1968) described a case where piston cores taken from the shelf off Surinam contained disturbances in the lower 40 percent of each sample. The disturbance was caused, they believe, by an upward motion of the piston during pulling up, leading to a sucking up of sediment in the core.

Sampling with Shelby tubes during drilling operations may incorporate materials which have fallen into the hole during reentry. Radiography is useful in recognizing such disturbed material, preventing it from being incorporated into subsequent testing.

In some cases, underwater core sampling may bring up materials which are so soft that they cannot be extruded from the core cylinder without being utterly destroyed. In such cases, radiography of the unopened tubes is an essential part of their evaluation.

Chmelik, Bouma, and Bryant (1968) reported an interesting case in

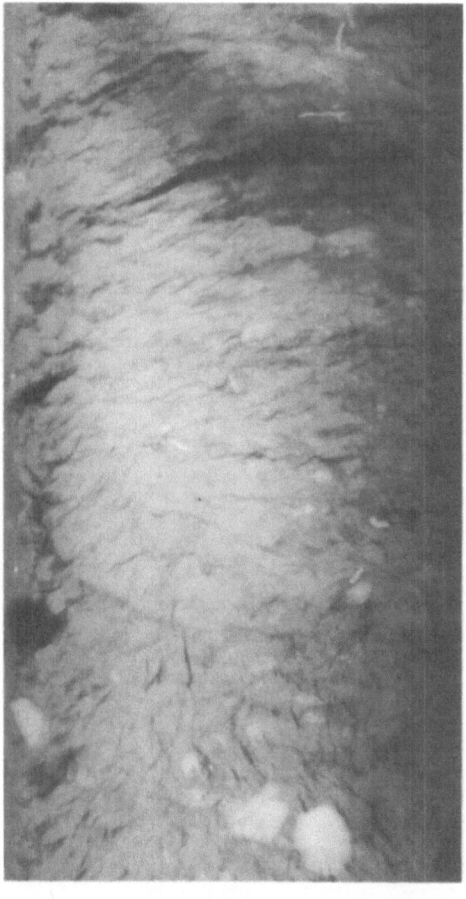

I INCH

Fig. 7-14. Expansion cracks caused by H_2S gas. Gulf of Mexico. Detail from split-core radiography. (Courtesy A. H. Bouma.)

which they believe their coring in a marine environment revealed an undisturbed void space or fluid zone within the sedimentary section. The void space, they believe, was not produced by sampling and may have been an exceptionally underconsolidated layer between two overconsolidated layers. Or it may have developed from a local electroosmotic condition. In any event, radiography of the unopened cylinder provided a unique means for evaluating this condition.

BIBLIOGRAPHY

Bouma, A. H. (1968). Distribution of minor structures in Gulf of Mexico sediments, *Trans. Gulf Coast Assoc. Geol. Soc.* **18**: 26–33.

———— and J. A. K. Boerma (1968). Vertical disturbances in piston cores, *Marine Geology* **6**: 231–241.

Chmelik, F. B., A. H. Bouma, and W. R. Bryant (1968). Influence of sampling on geological interpretation, *Trans. Gulf Coast Assoc. Geol. Soc.* **18**: 256–263.

Haase, M. C. (1967). *X-Radiography of Unopened Soil Cores*, MP 3–918, Corps of Engineers, Waterways Experiment Station, Vicksburg, Miss., 23 pp.

Rukavina, N. A. (1967). Rapid inspection of soft sediment cores by x-radiography, *Proc. Tenth Conf. on Great Lakes Res.*, Univ. Toronto, pp. 143–148.

Stanley, D. J. and L. R. Blanchard (1967). Scanning of long unsplit cores by x-radiography, *Deep Sea Research* **14**: 379–380.

———— and Gilbert Kelling (1967). Sedimentation patterns in the Wilmington submarine canyon area, *in: Ocean Sciences and Engineering of the Atlantic Shelf*, Marine Technology Society, pp. 127–142.

Chapter 8

PALEONTOLOGY

Applications of radiography in paleontology have been recognized and reported almost from the time of Roentgen's early experiments (Schmidt, 1948; Hamblin and Van Sant, 1963). Figure 8-1, taken from Roger (1947), shows one of its principal areas of application. Note the comparison between a photograph and a radiograph of a specimen of *Furcaster paleozoicus* Sturtz from the Lower Devonian in Germany. The radiograph reveals a much more complete picture of the specimen. To obtain this picture in a conventional way would have required several hours of scraping and might have resulted in ruining the specimen.

Success in radiography of the above sort is dependent on several factors. The fossil material must be of a density which is different from that of the matrix and sizeable enough not to be obliterated by a thick matrix. The matrix can, of course, be trimmed away in order to enhance a view of the fossil material. Zangerl (1965) provides a simple chart which shows some general rela-

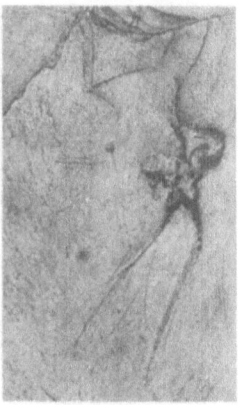
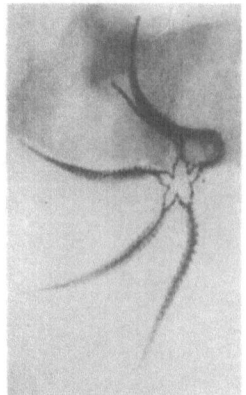

PHOTOGRAPH   RADIOGRAPH

Fig. 8-1. Comparison between photograph and radiograph of *Furcaster paleozoicus* Sturtz, Lower Devonian of Germany (after Roger, 1947).

tions between material and thickness of fossil and matrix and the potential quality of radiographic registration. In practice, these relationships are obtained by trial and error.

Radiography is exceptionally useful in revealing the undisturbed positions of shells and shell fragments (Fig. 8-2), root patterns and burrows (Fig. 8-3), and other remains in the original sediments. Population counts, burrowing depths, positions in relation to sedimentary structure, etc., can be evaluated without the tedious cutting away of cores or box samples and the destruction of the material which is necessary when using other methods.

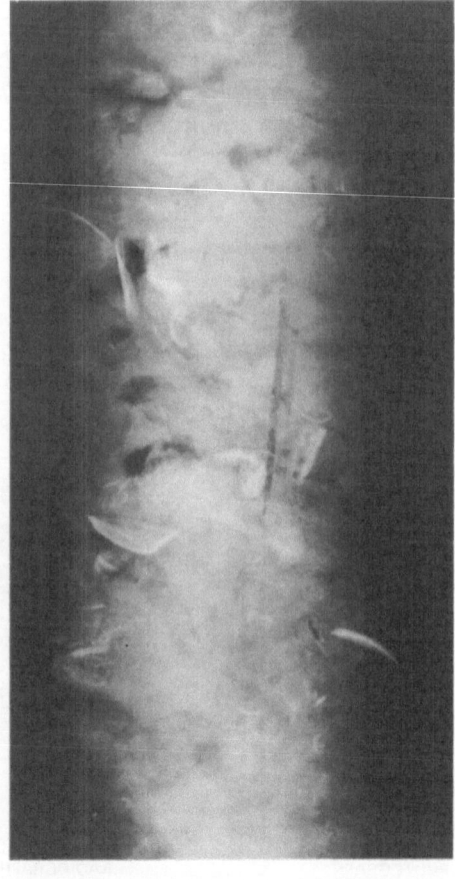

I INCH

Fig. 8-2. Shells and shell fragments seen in a split-core radiograph from bottom sediments in the Gulf of Mexico. (Courtesy A. H. Bouma.)

L INCH

Fig. 8-3. Patterns of large and small burrows seen in a split-core radiograph from bottom sediments in the Gulf of Mexico. (Courtesy A. H. Bouma.)

Stereoradiography of fossil remains is also practicable (see Figs. 3-9 and 3-10). Peyer (1934) reported favorably on the uses of radiography for examining rock-enclosed vertebrate remains prior to making decisions on extracting the material.

Howard and Henry (1967) and Howard (1968) observed bioturbations and the burrowing of invertebrates in aquaria by means of X-ray examination. Howard pursued these studies by means of time lapse exposures. By these means, the process of burrowing could be followed and burrowing rates could be determined.

Figure 8-4 illustrates the advantage of radiography for examining in-

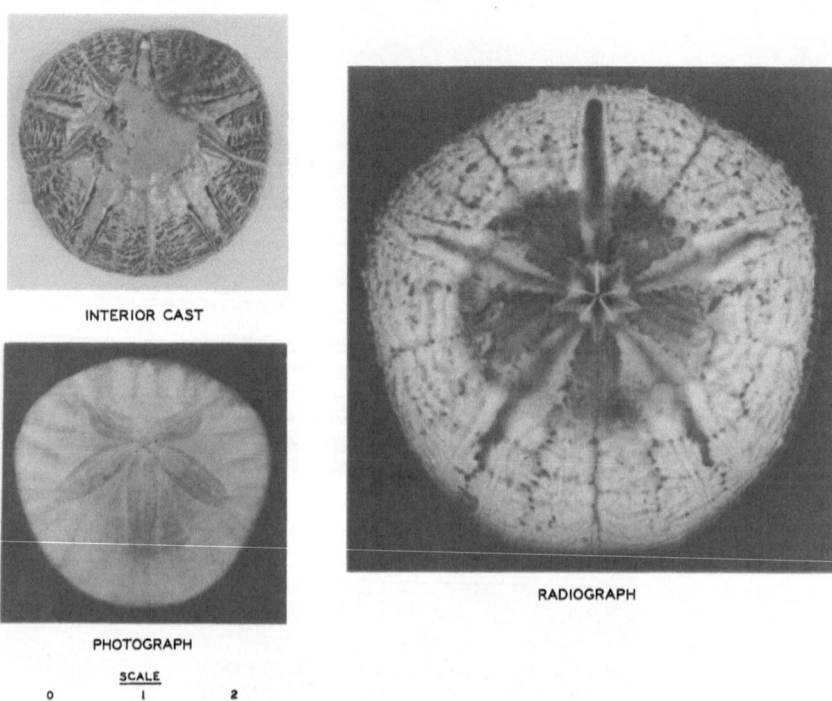

INTERIOR CAST

RADIOGRAPH

PHOTOGRAPH

SCALE

0   2

INCHES

Fig. 8-4. Comparison between specimen, interior cast made of Wood's metal, and radiograph of *Dendraster excentricus*. (Courtesy James A. Wolleben.)

ternal structure in fossils. The illustrations are from studies being made by James A. Wolleben on the sand dollar *Dendraster excentricus* from the Holocene. The interior cast was made by pouring Wood's metal into the test, allowing it to harden, and then dissolving the calcite away with HCl. The replica of the canals and passageways of the test interior then serves as an aid in interpreting the radiograph. The radiograph, however, contains much fuller detail of the interior of the specimen than was obtained in the replica. Again, with radiography, the specimen is not destroyed and much detail is registered which would be tedious and difficult to get by other means.

Microradiography, the technology of which is discussed in Chapter 10, has been found to serve effectively for micropaleontological studies. Microradiographic techniques as applied in the study of microfossils have been described by Hedley (1957) and Hooper (1959, 1960, and 1965). Schmidt (1952) described an adaptation which can be made with x-ray diffraction equipment in order to produce contact microradiographs of microfossils. She removed the special camera normally attached to a window of the diffraction unit. The removal of the camera made it possible to use the x rays without any modification of the x-ray diffraction apparatus itself. The diffraction camera was adapted by mounting an unexposed Eastman spectroscopic plate of extremely fine-grained emulsion on which the microfossil had been placed with gum tragacanth. The x rays were passed into this chamber through a piece of black paper which covered the camera chamber, the unit producing soft radiation of any one of several wavelengths depending on the target used. No filtering of the radiation was necessary. For a microfossil of about 0.5 mm in thickness, exposures with a copper target source were 25–35 kV, rarely 45 kV, 15–20 mA and 5–20 min. Distance from target to film was 7 in.

Hooper (1959) also made contact microradiographs of microfossils using an adapted x-ray diffraction unit. Registration was on a Kodak Maximum Resolution Plate. Specimens of foraminifera were affixed to "sellotape" which was then placed with the nonadhesive side against the emulsion side of the plate. Exposure times varied from 5 h at 18 mA and 30 kV to 7 h at 18 mA and 30 kV. The lengthy exposure time was compensated for by the large number of specimens which could be handled at one time. Some exposures included as many as 200 specimens on a single plate. Also, two cameras were operated from one x-ray tube concurrently. Though contact microradiography is normally done in a vacuum in order to get optimum results with very soft radiation, radiographs of calcareous tests on fossil foraminifera could be made effectively without using a vacuum.

According to Hooper (1965) projection microradiography (see Chapter 10 for its technology) provides a greater penetrating power than does contact microradiography. Resolution is approximately equal to the focal spot diameter or about 0.1 μ. The greater depth of focus is useful for more effective stereomicroradiography. Geometric distortions associated with projection may occur, much more so than with contact microradiography; however, corrections can be made for geometric factors.

Contact microradiographs can be viewed by examining them through an optical microscope or by enlarging their image through projection. McLung (1964) suggests that high-resolution plates permit practical magnifications up to 500×. Kodak Type M radiographic film, used in this manner, may be viewed successfully at magnifications up to about 60×.

Figure 8-5 shows the type of measurements that could be made on the contact microradiographic image of an operculine specimen. Hooper (1959) first took the radiograph and enlarged it 100× by projection in order to trace the diagram. Measurements were then made on the tracing with a ruler and protractor. Measurements taken were as follows:

1. Test diameter. a) North–South direction. b) East–West direction.
2. Number of whorls.

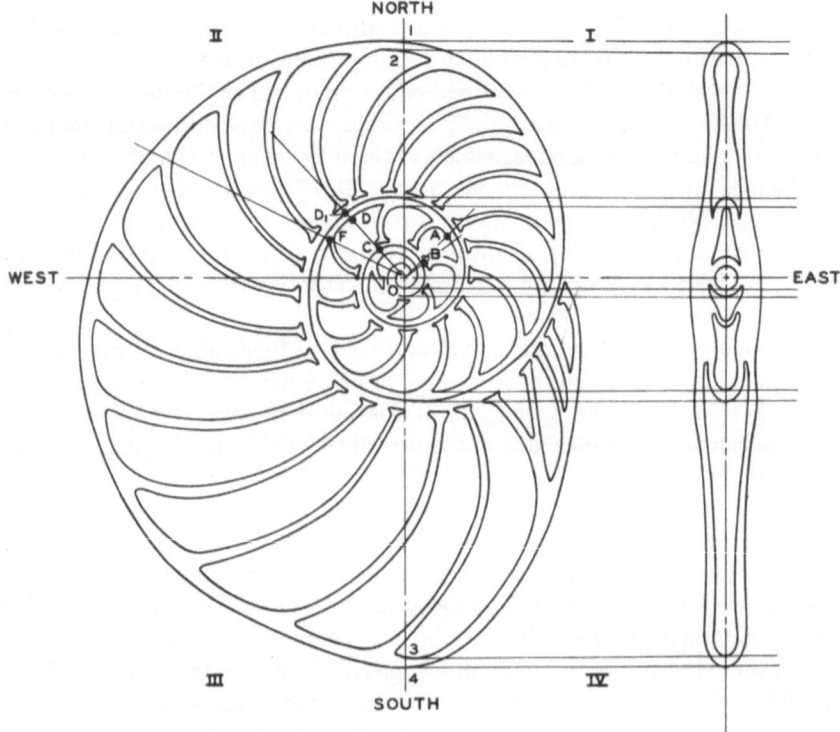

Fig. 8-5. Diagram showing measurements made on radiographic film of an operculine specimen (after Hooper, 1959).

3. Angle OAB.
4. Angle DOF.
5. Surface of the median section of first three chambers. a) Proloculus. b) Second chamber. c) Third chamber.
6. Proloculus diameters. a) Internal North–South. b) Internal East–West. c) External North–South. d) External East–West.
7. Second chamber, lengths. a) AB. b) CD. c) CD_1.
8. Whorl heights for 1st, 2nd, 3rd, and 4th whorls at each quarter whorl.
9. Number of chambers to fixed distances measured around the whorl (excluding proloculus).
10. Whorl heights to fixed distances around the whorls. a) To 3 mm. b) To 6 mm. c) To 12 mm.
11. The measurements a) N_2S_3 and b) N_1N_2.
12. Diameter of test at N_1S_4.

13. Maximum septal filament length per whorl. a) First whorl. b) Second whorl. c) Third whorl.

Hooper believes that the maximum combined error from penumbral unsharpness, granularity of the photographic emulsion, geometric effects, distortion from projection, and pencil error in tracing the projection is on the order of 10%. Most of the time, he believes the actual error is less. In comparison, the possibility for error by conventional sectioning and direct measurement, with loss of the specimen, is 7%.

Thus, radiographic experimentation in paleontology has had a long history and there have been numerous demonstrations of its usefulness. Yet, as Zangerl (1965) and Schmidt (1948) have shown, not many paleontologists have adopted the technique for use as a standard laboratory procedure. Part of the reason, Zangerl observes, is from the unfamiliarity of paleontologists with radiographic equipment and with procedures for processing film and prints. Another difficulty is that new ways must be learned for reading the shadow pictures which are produced by radiography.

BIBLIOGRAPHY

Hamblin, W. K. and J. Van Sant (1963). Radiography in paleontology, *in:* George L. Clark (ed), *The Encyclopedia of X-Rays and Gamma Rays*, Reinhold, New York, pp. 684–686.

Hedley, R. H. (1957). Microradiography applied to the study of foraminifera, *Micropaleontology*, 3 (1): 19–28.

Hooper, Kenneth (1959). X-Ray absorption techniques applied to statistical studies of foraminifera populations, *Jour. Paleon.*, 33 (4): 631–640.

—— (1960). Microradiography in quantitative micropaleontology techniques, *in:* A. Engström, V. Cosslett, and H. Pattee (eds), *X-Ray Microscopy and X-Ray Microanalysis*, Elsevier, New York, pp. 216–223.

—— (1965). X-Ray microscopy in morphological studies of microfossils, *in:* B. Kummel and D. Raup (eds), *Handbook of Paleontological Techniques*, Freeman, San Francisco, pp. 320–326.

Howard, J. D. (1968), X-ray radiography for examination of burrowing in sediment by marine invertebrate organisms, *Sedimentology*, 2 (3/4): 249–258.

Howard, J. D., and V. J. Henry (1967), Use of x-ray radiography in the study of bioturbate textures. 7th, Intern. Sed. Congr., Reading, 4 pp.

McLung, R. W. (1964). Studies in contact radiography, *Materials Res. & Standards*, pp. 66–69.

Peyer, B. (1934). Über die Röntgenuntersuchung von Fossilien, hauptsächlich von Vertebraten, *Acta Radiologica* 15: 364–379.

Roger, J. (1947). Résultats fournis par l'application des rayons x a la paléontologie, *Bull. Soc. Géol. France* 17 (5): 483–491.

Schmidt, Ruth A. M. (1948). Radiographic methods in paleontology: a progress report, *Am. Jour. Sci.* 216: 615–627.

—— (1952). Microradiography of microfossils with x-ray diffraction equipment, *Science* 115: 94–95.

Zangerl, Rainer (1965). Radiographic techniques, *in:* B. Kummel and D. Raup (eds), *Handbook of Paleontological Techniques*, edited by Freeman, San Francisco, pp. 305–320.

Chapter 9

SOIL MECHANICS

The first application of radiography to gathering primary data in soil mechanics can be credited to Emil Gerber (1929) as part of a doctoral dissertation at Zurich. His testing was described and illustrated by Tschebotarioff (1950) in a discussion of related work done by Berdan and Bernhard (1950). Essentially, Gerber succeeded in tracing the movement of lead pellets in a sand that was subjected to plate loading. His film imagery was not clear enough to be reproduced but it served for measurements of pellet displacements. Apparently no further work was done until Davis and Woodward (1949) reported tests with footings on sands. In the same year, J. J. O'Dea, Jr., used radiography in studies of vibratory characteristics of sands. Shortly afterwards, Berdan and Bernhard (1950) applied radiography in compaction studies to analyze soil densification processes in silt and in sand. Bergfelt (1956) traced lead-shot displacements with footing tests on blocks of clay, and Watkins (1957) used radiography to trace the movements of lead shot around models of flexible culverts buried in loess.

Since then, there has been a strong and continually increasing development in the application of radiography in soil studies. Experimental work has been done on soil deformation under conditions of plane strain, displacement and compaction of soils around piles, footing displacements in clays under static and dynamic loading, blast effects on clays, interpretation of triaxial tests, and other laboratory-oriented research. Imaginative tests using flash x-ray techniques have been applied in soil-dynamics research. Compressive wave propagation through soils has been recorded and work has been done on monitoring the development of explosion cavities, cratering processes, the interactions of soils and structures during shock loading, and projectile penetration into earth materials. This chapter will attempt to assess the development of applications in these and related areas.

EFFECTS OF X RAYS ON SOILS

One of the great merits of x radiography is that it is nondestructive. At least it is generally assumed to be nondestructive though x rays are known to be harmful to living tissues and radiation has been used to sterilize soils of their microorganisms (cf. McLaren *et al.*, 1962). Is the assumption that radiation does not affect soils a safe one to make for those soils where very delicate measurements of properties are to be obtained? Leitch and Yong (1967) recorded some observations that are relevant to this question.

Clark (1955) indicated that x rays can cause flocculation in clay slips by the breakdown of certain molecules into their component ions, at least temporarily. Thus, radiation can upset the dynamic chemical equilibrium of hydrogen ions and hydroxyl radical groups present in pore fluids and cause a temporary increase in the production of ionic components. These ionic components might become permanently associated with the clay particles thus altering the properties of the clay soil.

Leitch and Yong observed this effect by radiating suspensions of calcium–kaolin and calcium–attapulgite as they were undergoing settlement in cylinders. Their system, involving a vacuum sedimentation unit with a

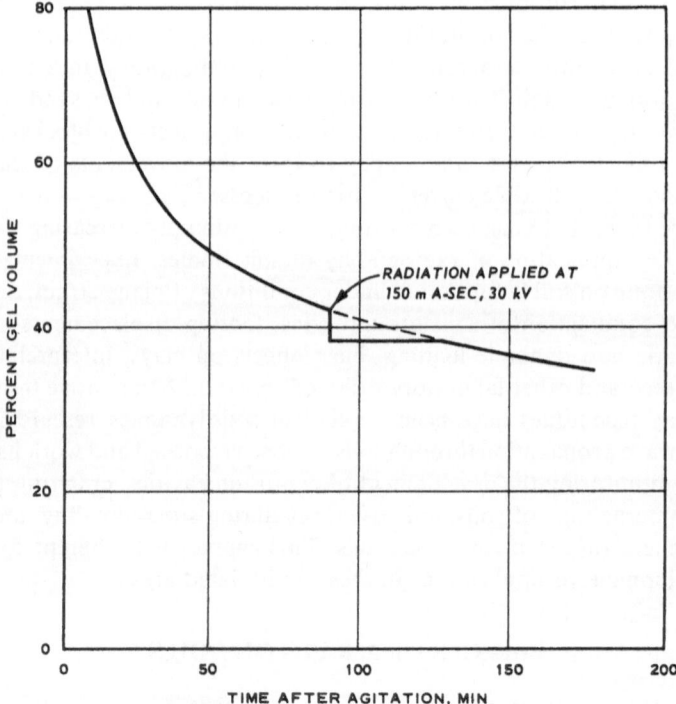

Fig. 9-1. Effect of radiation on the settlement of a calcium–kaolin suspension (after Leitch and Young, 1967).

consolidation cylinder, was described in detail by Yong, Japp, and Leitch (1964). Sedimentation rates were observed under normal conditions and with exposure to radiographic dosage initiated after the sedimentation process had progressed for 90 min. Figure, 9-1 shows their observation of the effect of 150 mA-sec at 30 kV on a calcium–kaolin suspension. The introduction of radiant energy has an immediate effect on the sedimentation curve followed by a slow adjustment in which the irradiated curve approaches the reference curve. The suggestion is that there is a radiation-induced alteration of fabric, at least on a temporary basis.

The effects of radiation on the determination of Atterberg limits are shown in Table 9-1. The results suggest that there may be a radiation induced effect on fabric where there are relatively high water contents. However, the magnitude of the effect is of a low order and at least some of the variation can be due to experimental error. Leitch and Yong suggest the probability that all the variation in experimental error is less than 50%. The authors note also that the soil was allowed to remain undisturbed for 15 min after radiation and before the standard Atterberg test procedures were initiated. They suggest that though the effects measured may be time dependent they are not necessarily entirely transient.

Leitch and Yong then examined the effects of radiation on the measured residual shear strength of clays subjected to reshearing. Resheared samples, exhibiting minimum shear strength, were exposed to radiation intensities of 75 mA-sec at 25 kV, for attapulgite, and 150 mA-sec at 30 kV, for kaolin, and allowed to stand for 15 min between radiation and initiation of reshearing. The 15-min period was intended to parallel the time lapse followed in making the Atterberg determinations. The sample was resheared and the resulting

Table 9-1. The Effects of X Radiation on the Atterberg Limits of Clay Soils (after Leitch and Yong, 1967)

Soil type	Pore salt	X-ray dose	Liquid limit	Plastic limit	Shrinkage limit
Attapulgite	Na_3PO_4	75 mA-sec, 25 kV	171.8	149.6	81.4
		Nil	170.8	144.9	80.2
	Nil	75 mA-sec, 25 kV	174.0	145.3	79.5
		Nil	172.0	145.1	72.5
	$CaCl_2$	75 mA-sec, 25 kV	177.0	144.6	78.1
		Nil	175.0	150.7	77.8
Kaolin	Na_3PO_4	150 mA-sec, 30 kV	52.8	36.7	27.6
		Nil	52.7	33.6	37.7
	Nil	150 mA-sec, 30 kV	51.8	38.2	37.6
		Nil	51.0	37.3	37.7
	$CaCl_2$	150 mA-sec, 30 kV	52.1	38.7	35.4
		Nil	52.0	39.0	34.9

stress–strain curves were compared with the residual stress–strain curves for the same material in an unradiated condition. Within the limits of accuracy of the apparatus used, the curves were found to be identical for radiated and nonradiated samples. Though there is no measured effect on shear strength, there is a possibility of soil-fabric alteration due to the radiographic process, though it may be of a low order except in those cases where there is a condition of high moisture.

Bloedow (1962) considered the effects of radiation on laboratory instrumentation. He concluded that conventional instrumentation, consisting of accelerometers, stress and strain gages, pressure gages, and electronic semiconductor components, are not significantly affected by radiation, nor would 250 keV provide enough energy for a detectable change in transducer properties. Piezoelectric devices operate on the basis of electrical signals produced by mechanical deformation of polarized crystalline devices. Radiation sufficient to produce ionization can produce noise in the electrical field of these devices. However, Bloedow felt that these sensors too would not be adversely affected by the radiation used in soils tests.

SOIL DEFORMATION PATTERNS IN MODEL TESTS

The applications of radiography to model tests may take a variety of forms, but they nearly always include the burial of lead markers in a specimen so that the subsequent movements of these markers can be traced. The model may be made of small dimensions so that x rays are passed through the entire model, either during various stages of testing, continuously, or at the termination of testing. Or a test may be run on a large model with a section of the model cut out and observed after testing. If the model is of remolded materials, lead pellets can be placed on the surfaces of lifts as the model is built up. With undisturbed soils, small incisions or needle injections can be made for the introduction of pellets.

An example of after-testing radiographic examination of a cross section made from a large model is shown in Figs. 9-2 to 9-4. The full section of the model is 1 ft deep and 2 ft wide. The model is of remolded backswamp clay compacted in four horizontal lifts, each about 3 in. thick. Lead pellets were positioned on the surface of each lift. After static loading of the plate as indicated in Figs. 9-2 and 9-3, a full section was cut out as a brick mosaic. Each brick was sliced to a thickness of $\frac{3}{8}$ in., photographed, and radiographed. The pictures were then reassembled as mosaics. This work was done at the Waterways Experiment Station in the Soils Dynamics Branch and in the Radiological Laboratory. Similar methods of preparation were reported by E. B. Perry (1968a and 1968b) in tests of small-scale footings in clay. The photograph mosaic in Fig. 9-2 illustrates the difficulty, in this case the impos-

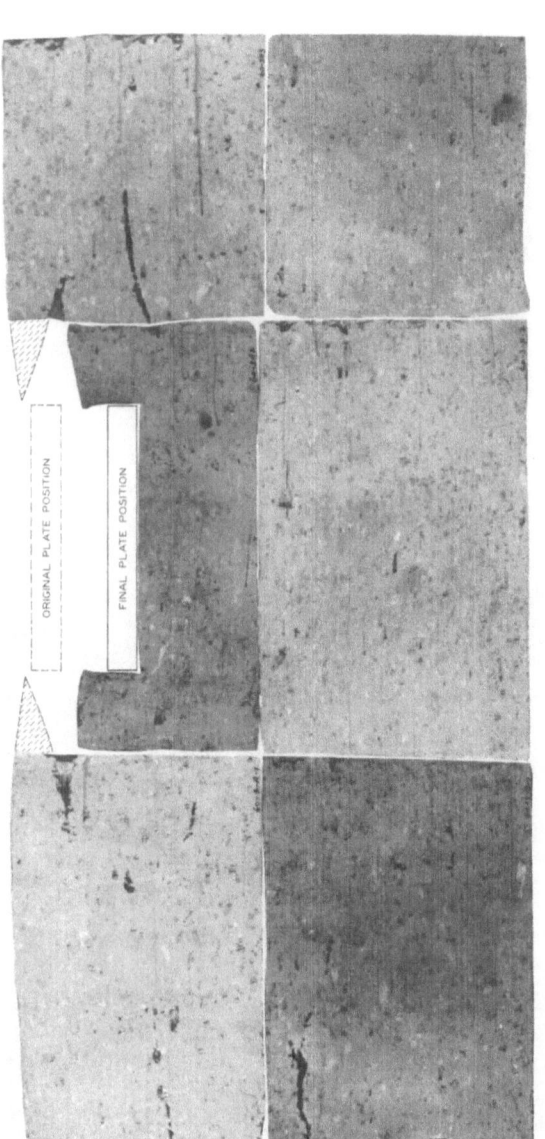

PHOTOGRAPH MOSAIC

STATIC TEST 60-3

P_{MAX} = 4.86 KIPS
Z_{MAX} = 2.14 IN.
t_O = 103 MIN

SCALE IN INCHES

ORIGINAL PLATE POSITION

FINAL PLATE POSITION

Fig. 9-2. Photographs in mosaic section of remolded backswamp clay after testing by static-plate loading.

RADIOGRAPH MOSAIC

STATIC TEST 60-3

$P_{MAX} = 4.86$ KIPS
$Z_{MAX} = 2.14$ IN.
$t_O = 103$ MIN

SCALE IN INCHES

Fig. 9-3. Radiographs in mosaic section of remolded backswamp clay shown in Fig. 9-2. Note local shear patterns and positions of lead pellets.

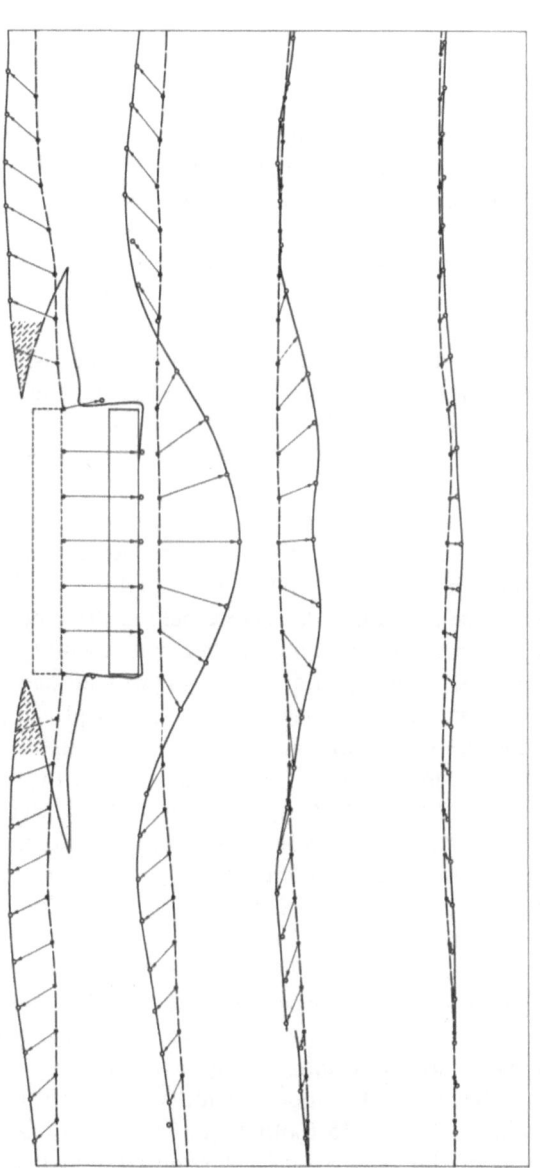

DISPLACEMENT VECTORS
STATIC TEST 60-3

P_{MAX} = 4.86 KIPS
Z_{MAX} = 2.14 IN.
t_O = 103 MIN

SCALE IN INCHES

Fig. 9-4. Displacement vectors interpreted for movement of lead pellets seen in Fig. 9-3.

sibility, of recognizing the disturbances induced in a remolded clay. Even the lifts cannot be traced. Figure 9-3, in contrast, not only shows the four lifts and their varying amounts of deformation, but also shows an intricate pattern of discontinuous shear surfaces related to the final plate position. Figure 9-4 shows a vector analysis based on inferred pretest positions of the lead pellets. As there are no pretest radiographs, the original pellet positions are based on a careful geometric positioning of the pellets during construction of the model. The $\frac{3}{8}$-in. slices that were radiographed were registered with a focal distance of 18 in. Thus, distortion by radiography was minimal and could be ignored in this analysis. Exposure was for 45 sec at 18 mA and 25 kV using Kodak Type M Ready Pack Industrial x-ray film. Radiation was with a Philips Industrial x-ray unit with a 100-kV beryllium window tube. Besides showing the feasibility of a mosaic-block, thin-slice, radiographic method of analysis for these tests, the results showed that it is possible to classify the failure mechanisms (i.e., general shear, local shear, punching shear, etc.) for both static and dynamic small-scale footing tests in clay.

Earlier work by Davis and Woodward (1949) showed the applicability of x-ray examination of footings in models of granular soils. Their laboratory setup had the model and x-ray tube in a shielded booth, with the operation being viewed through a leaded glass window. The sidewalls of their model were 3 or $4\frac{1}{2}$ in. apart. The granular soils that formed the test specimens were marked with lead pellets. Film was placed behind the model. In other tests, a fluorescent screen was used, photographed with a 35-mm camera. Photographs were made periodically after increments of load had been applied to the footings and projected in order to measure displacements. Corrections were made for the effects of parallax. The radiographs were successful for tracing internal movements in the soils; however, the small size of the models made correlation with large-scale behavior uncertain. Their radiographs indicated patterns of displacements in the sands through movement of the lead pellets and density differences along failure surfaces. They further observed that a densified zone of triangular section may develop beneath a footing and that the wedge may then move as if it were part of the footing. Also, shearing failure may develop with characteristic patterns on one or both sides of the footing. Displacements were inappreciable at depths below 3 footing widths. The failure zone intersected the surface at a distance from the footing of not more than 5 footing widths and lateral deformation was inappreciable beyond 6 footing widths. Thus, they concluded that a model should be at least 15 footing widths wide in order to provide a free development of the failure zone. They also found that the friction between the soil and the wall of the container influenced the shape of the failure surfaces for a distance from the sidewall equal to about two footing widths, and that the shape of the failure zone was closely approximated on a vertical plane by a trace with the shape of a logarithmic spiral.

In many tests, the footings settled unevenly, implying that the models were minutely nonuniform in original density and that there were irregularities in degree of confinement during shearing action. Accessory tests, both with direct shear and triaxial compression, showed that sandy soils of medium to high density and under moderate confining pressures had angles of internal friction on the order of 36°; on loose soil and small confining pressure, the angles of internal friction were as low as 15°. The authors suggest that in bearing tests on loose soils, where an increase in density takes place near the footing when load is applied, the resulting variation in shear resistance is an important factor to consider in estimating bearing capacity. Thus, in small-scale testing, particularly in laboratory models, the control of uniform density of the test specimen is of great importance.

X-ray studies of plate loading on blocks of undisturbed clay were reported by Allan Bergfelt in 1956. Bergfelt ran his tests on 35-cm cubes of Gothenberg clay to obtain comparisons with field tests run in test pits. His cubes of clay were set in paraffin-lified wooden boxes. Lead pellets were pushed into the clay using special needles with clips to measure exact placing of the pellets. Spherical pellets were used in the center, cubic ones in a vertical line at the edge of the plate. Insertion of the pellets did not appear to affect the model. The arrangement for x raying involved a film cassette on one side of the box and an x-ray tube on the other. The x-ray tube was at a distance of 68.2 cm from the center of the 35-cm model. The film cassette was 20.2 cm from the center of the model. Thus the setup produced a large conical distortion. Lead pellets near the surface were difficult to distinguish. To overcome this, a test was made with the x-ray tube level with the top of the model, but this turned out to be worse because of scatter and undercutting.

The x-ray apparatus was used at its maximum capacity. Exposures were made in five periods of 4.5 sec each. A 2-min interruption was made between each period. Intensity was 125 kV and 60 mA.

Failure load in the model was only 10 to 15 % less than in field tests despite disturbances in excavating the blocks, handling, and inserting pellets. However, the size of plates and loading speed were less than in field tests. Deformations in the interior of the clay were greatest in a deep wedge under the plate, but no slip surfaces were noticed even with settlements as great as 10 % of the plate breadth. The author suggested that slip surfaces might have been apparent if the test measurements had been more exact.

In 1949, James J. O'Dea, Jr., made a study of the vibratory characteristics of loose sand (bulk specific gravity of 1.7) in a sand box $14 \times 14 \times 21$ in. The model was made with 2-in. layers of sand separated by 1-in. layers of lead pellets. A pile ($1\frac{1}{8}$ in. in diameter and 24 in. long) was sunk into the sand by vibration. The pile was removed and reintroduced by vibration several times in order to fully compact the sand within the pile area. Settlement of the sand was noted from the displacement of the layers of lead shot. The study is a landmark as it was the first time that gamma rays, emitted from radium, were used for radiography of a soil model.

O'Dea obtained his radiation from a 100-mg capsule of radium. Registration was on Ansco Hi-Speed x-ray film (14 × 17 in.) between lead-foil intensifying screens. The film cassette was placed on one side of the model and the radium was placed 25 in. away on the other side. Exposures were made before and after testing. Exposure time was 48 h. A portion of one of the radiographs is shown in Fig. 9-5. Note the depression of the lead-pellet layers along the sides of the pile and at the bottom. The radiographs were not satisfactory for determining changes in sand density that resulted from com-

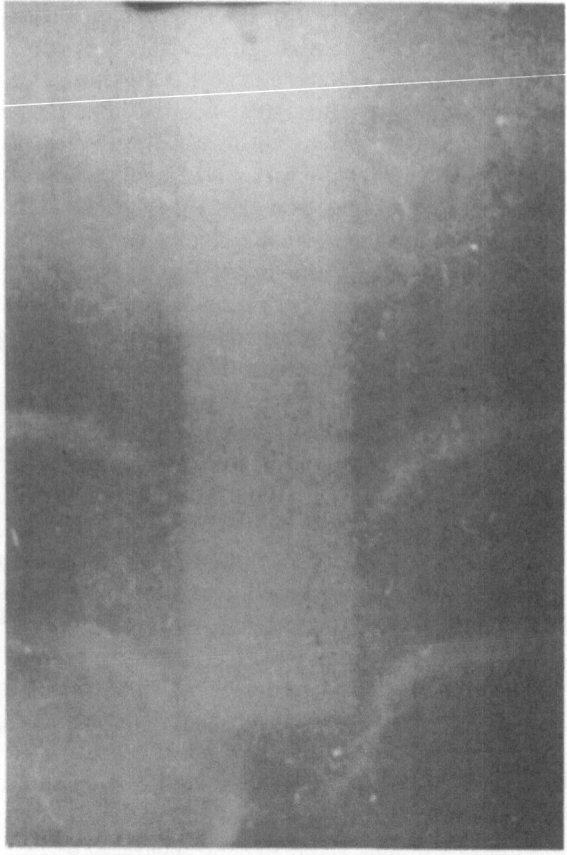

Fig. 9-5. Radiograph of pile model in sand made with gamma radiation from a radium source. Note patterns of lead pellets around the pile. (Courtesy James J. O'Dea, Jr.)

paction, although the patterns of settlement could be examined with reasonable accuracy. The long exposure is accompanied by a fogging of the film by scatter and the short focal distance contributes to distortion of the image. However, O'Dea was able to determine that a hemispherical volume of soil was compacted beneath the driven pile and that soil surrounding the pile sloped towards the pile at approximately 40°, or the angle of repose of the sand–pellet mixture.

More recently, radiography with a cobalt-60 source was used by Robinsky (1963) and Robinsky and Morrison (1964) to study the displacement and compaction of sand around instrumented model piles. Their cobalt-60 source had a diameter of 0.25 in. and an intensity of 120 effective curies. The test box, filled with Ottawa sand, had a minimum interior thickness for radiographic penetration of 19.75 in. The walls of the box, of acrylic plastic, presented a thickness of 1.75 in. to radiation. The box was 19.75 in. thick, 28 in. wide, and 32 in. high, with a volume of 10.2 ft. The source-to-film distance was 7 ft 5¼ in. and the object-to-film distance was 10.87 in. The diameter of lead shot was 0.13 in. Kodak Industrial x-ray film type AA was used. Two

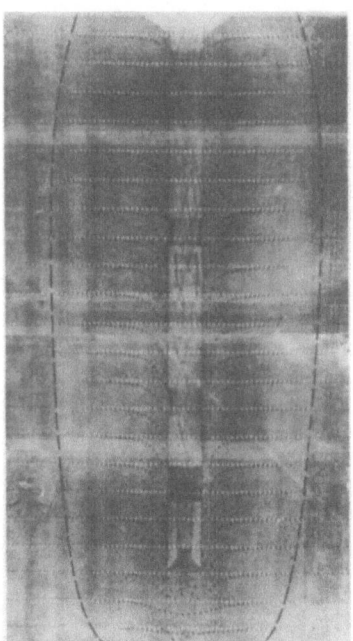 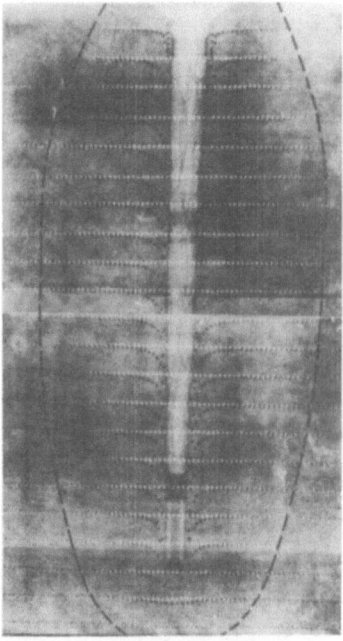

Fig. 9-6. Radiographs made with a cobalt-60 source. Left, straight pile with rough surface in loose sand; right, tapered pile with rough surface in loose sand. Length of piles: 20 in. Note before and after positions of lead pellets. (Courtesy E. I. Robinsky.)

sheets, each 14×17 in. provided a 28×17 in. area for exposure. Lead screens 0.01 in. thick were used on both sides of the film.

Exposures were made before and after testing. Exposure times were 15 to 20 h for standard Ottawa sand (ASTM Designation C 109) at $D_r = 17\%$, and 20 to 23 h for standard Ottawa sand at $D_r = 37\%$.

Piles that were straight and tapered, rough and polished, were driven into the sand to maximum penetrations of 20 in. The piles were instrumented internally with strain gauges to determine load distribution to the surrounding soil through the pile walls and the pile points.

Figures 9-6 and 9-7 show examples of interpretative radiographic prints

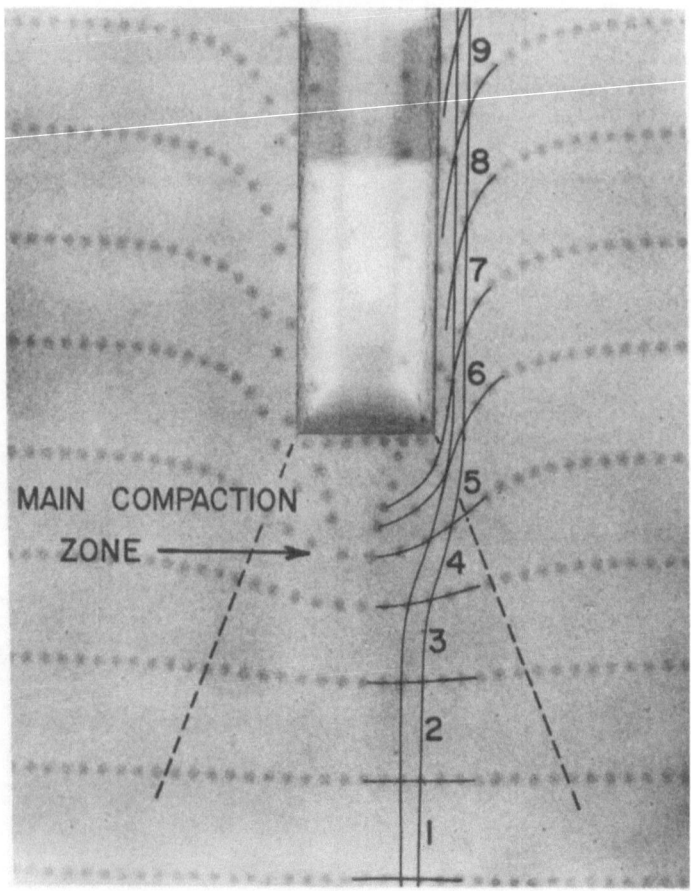

MAIN COMPACTION
ZONE ──────────▶

Fig. 9-7. Detail of displacements near base of straight pile in Fig. 9-6. (Courtesy E. I. Robinsky.)

that were prepared for the straight and tapered piles. These examples are for piles with rough surfaces and in loose sand. The white dots are the positions of lead shot prior to driving the piles. The black dots are positions after driving the piles. The plates were produced by superimposing the first radiograph onto a diapositive of the second radiograph. Where lead shot did not move in space and photographic densities were balanced, the reversal of the two radiographs resulted in partially obliterating the images of the lead shot. The envelopes shown by the dashed lines indicate the limits of recognizable displacements of the lead pellets. Figure 9-7 shows in detail the film record of only the displaced shot pellets near the base of the straight pile.

The general shape of the displacement envelope was similar for all tests. Figure 9-8 shows contours of interpreted equal relative densities that were calculated from pellet displacements within the overall envelope of disturbance. Compaction around both the straight-sided and tapered piles resulted in a seemingly erratic accumulation of high- and low-density areas. Load transfer is believed to have developed through numerous arching patterns.

Watkins (1957) applied the procedure of making sequential x-ray pictures of models containing cylinders buried in loess in which a grid of lead pellets served as markers. Exposures were made on separate sheets of film after each

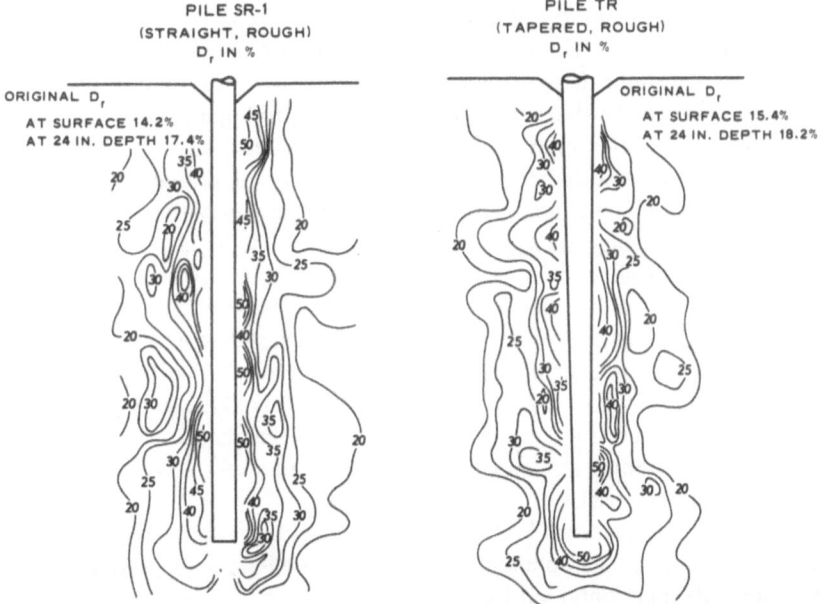

Fig. 9-8. Contours of equal relative densities around piles interpreted from pellet displacements observed in radiographs (after Robinsky and Morrison, 1964).

load increment was applied. By superimposing the radiographs, the sequential soil displacements could be traced by means of movements in the shot pattern. Watkins presents some striking diagrams of his results but did not report any of the details of his radiographic equipment, distance measurements, exposure times, or film types.

Soil-dynamics studies of models observed by flash radiography have been reported by Bloedow (1962), Baker and Janza (1966), and Gornak (1967).

Bloedow (1962) also used a technique whereby a dynamic model was subjected to steady radiation and the record was taken continuously on a sheet of film carried in a revolving drum and exposed through a slit. The soils used were Ottawa sand and a silty–sandy–clayey mix. Lead pellets were imbedded in the specimens. The specimens were mounted in a shock tube and loading pressures were 20, 60, and 80 psi. Peak pressure was applied against a thin membrane on the free surface of the soil for about 6 msec. The soil thickness observed by radiation was 5 in. Radiation was 300 kV at 10 mA. The metal slit was $\frac{1}{16}$ in. thick, film speed was 21 in./sec, Kodak Royal Blue Medical x-ray film was used, and the focal distance was 2 ft. Traces were registered of total pellet displacement and pellet velocity was calculated.

On a number of the film records there were measurable changes in film density, believed to have been a direct result of changes in soil density induced by the stress wave front. By calculation of postulated changes in mass absorption of the x rays (see Chapter 11 for a description of this method), it was possible to determine density changes across the stress wave front. It was impossible, however, to determine the velocity of propagation of this density change because of the relatively long exposures needed with the steady radiation source. Also, the degree of unsharpness of the pellet traces was such that strains in the soil could not be measured to the accuracy that would be required for calculating density changes on a volumetric basis.

Bloedow's flash x-ray tests were run on the same types of samples and with a simulation of the same loading. A Zenith flash x-ray machine was used with a capacity of 30 flash exposures per sec. Each flash had a maximum beam intensity of 4 mR at a distance of 1 m when a maximum voltage of 150 kV was used. Pictures were from an image-intensifier tube. Due to the time intervals between flashes, it was not possible to obtain a sequence of useful radiographs during a single test and the propagation of stress wave fronts in the soils could not be observed.

The more recent work by Baker and Janza (1966) considered flash x-ray equipment of greater sophistication and included tests of soils (silts and Ottawa sand) that contained buried cylinders made of aluminum tubes that were 4, 2, and 0.625 in. in diameter with wall thicknesses of 0.040 in. The soils, 2 to 6 in. thick, were also marked with lead pellets, but definition of the lead pellets was found to be poor. Lead cubes (0.1875 in.) were used with more success. Rectangular bars (0.1875 in. sq. and 0.50 in. long) were also used, with the long axis along the path of the x rays. However, the cubes and bars were difficult to place so that the edges would be parallel to the beams. Rotation of the pellets was found to distort the shadowgraphs in such a way that erroneous interpretations of translation were possible.

Loading of the models was with a hydraulic ram piston hitting the surface of the soil. The cylinders were both deformed and failed with peak loads of 100 psi.

Pretest experiments showed that sharper pictures were made by direct exposure on x-ray films as compared with the pictures obtained by a 35-mm high-speed motion camera placed behind an image intensifier.

The x-ray source was 24 to 28 in. from the recording surface. Clear Plexiglas supported by steel braces served as soil container walls; tests were also made with walls of laminated wood, which deformed excessively at pressures of 100 psi, and aluminum, 0.25 in. thick, which scattered and attenuated the x-ray beam excessively. Scattering of x rays in the soil and by the lateral walls of the container was high and a foggy background was produced on the image. Scatter was reduced by using a lead sheet with a window cut out to the size of the area that was to be examined. Thus, x rays were allowed to enter the soil only in the area of interest. Scatter was further reduced by using microline x-ray grids made by Liebel-Flarsheim Company of Cincinnati, Ohio. These grids are made of fine strips of lead, 60 to 150 strips per inch,

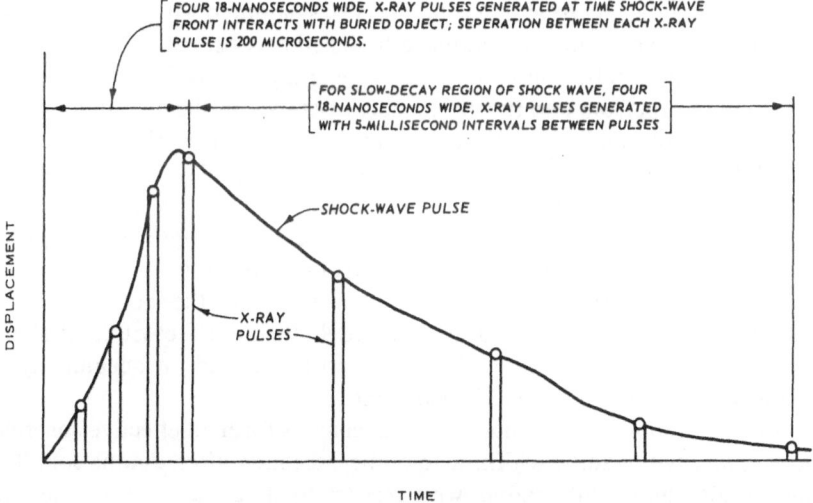

Fig. 9-9. Application of Field Emission Corporation flash x-ray system with variable pulse-repetition frequency (after Baker and Janza, 1966).

that clean scattered x rays out of the radiation that has passed through the soil sample. They are designed for specific focal lengths and must be selected to match the source-to-image distance.

Observations were made with a Zenith multiple flash x-ray system that generated eight 1-μsec pulses with a pulse-to-pulse time separation of 1000 μsec. Tests were also made with a Field Emission Corporation flash x-ray system that has the potential of producing 18-nsec pulses that are as close as 200 μsec apart. Possible variable programming of radiation bursts is shown in Fig. 9-8.

Baker and Janza suggest that the greatest advantage of the above tests is that they are successful for determining the interaction between soil and a structure under dynamic loading. In contrast, instrumenting such experiments with gages, especially at boundaries, is nearly impossible. Laboratory experiments on direct explosion coupling, cratering, and underground explosion cavities could be monitored by the high-speed x-ray techniques.

Experiments using a flash x-ray system to study ground shock and gas cavity expansion in models that simulated underground explosions were reported soon afterwards by Gornak (1967). Gornak used a remolded soil of sand and fines in a 7-in. cubic box (inside measure). made of ⅝ in.-thick plywood. Radiation was provided by a Field Emission Corporation Fexitron 730/2650 pulsed x-ray system. Minimum source-to-object length was 20.5 in. Registration was of single exposures on Kodak Medical x-ray film with Du-Pont Granex x-ray intensifying screens. The film cassette, 8×10 in., was placed inside an aluminum cassette, 0.25 in. thick, that in one case was placed flush against the soil cube and in other tests was mounted with an intervening air space. The explosions were induced by blasting caps. An ion probe, connected through a pulse source to a delay generator, served to sense arrival of the detonation front and to trigger the x-ray tube. As the detonation front reached the probe and closed the circuit to the pulse source, the pulse source sent a 90-v, 100-nsec pulse through the delay unit. The delay unit then started counting and at a preselected time sent a pulse to trigger the discharge of a capacitor bank to energize the x-ray tube. Time delay settings were computed from shock velocity information obtained from earlier tests.

The tests were found to be successful for determining the shock and cavity geometrics of the explosive charges. Figure 9-10 shows a cavity and shock growth after 27 μsec. There is also a ground shock and an accompanying ground surface motion of a vertical air shaft.

Gornak reported that small spherical charges form an effective spherical shock and cavity radius within a few microseconds after detonation. The shock radius and cavity radius with respect to time was measurable. His results further suggested that the final nuclear crater dimensions will be little affected by the venting of a one-third partially stemmed shaft.

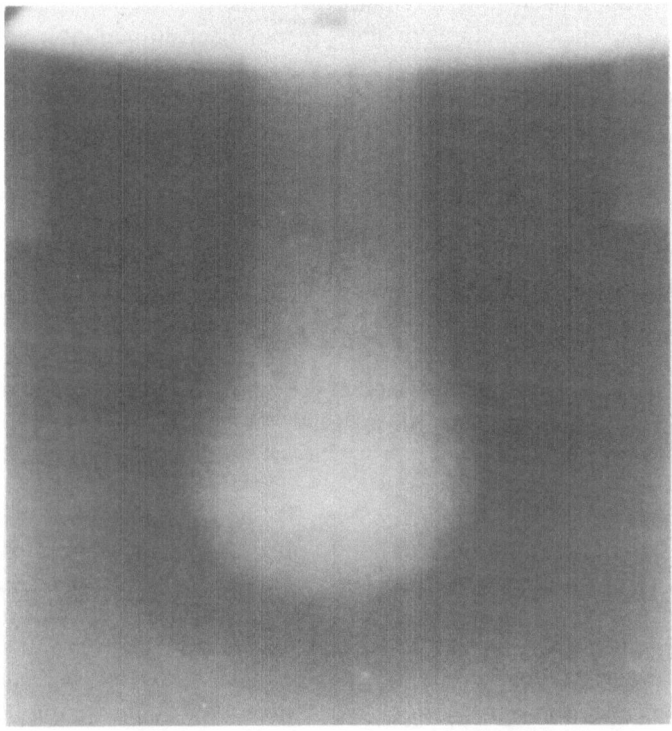

Fig. 9-10. Cavity and shock growth of a model explosion after 27 μsec. (Flash x-ray picture courtesy George Gornak.)

Colp (1968) reported a radiographic application in terradynamics involving the penetration of projectiles into natural earth materials. Colp's method was to excavate the material around an imbedded projectile, make a uniform slice of this material normal to the projectile, and radiograph the slice. A density determination was then made on a small portion of the material for a control value. Then an isodensity tracing (see Chapter 11) was made of the radiograph and density variations induced by the projectile were contoured.

PHYSICAL PROPERTIES OF SOILS

The physical properties of soils that can be interpreted successfully with radiography are:

1. Soil densities.

2. Inhomogeneities in remolded specimens.
3. Details of fractures or shear planes, depositional layering, secondary mineralization, plastic deformation, organic matter and fossils in undisturbed specimens.
4. Effects of details of soil structure on the interpretation of test results.
5. Effects of boundary conditions on the behavior of test specimens.

The use of radiography for measurement of soil density is of such considerable importance in soil mechanics that it is treated at length in Chapter 11.

The details of fractures, mineralogical inhomogeneities, and sedimentological features are discussed throughout this volume. However, the effects of these features on the strength properties of soils constitute a largely unexplored field.

Krinitzsky (1970) made some initial studies of such relationships on soils from the lower Mississippi Valley. A group of representative depositional environments were selected, including several prodelta and interdistributary deposits from the deltaic plain, buried Pleistocene material, backswamp deposits from the Atchafalaya Basin, clay plugs, point-bar topstratum, swales, and backswamp from the alluvial valley. All were at sites in Mississippi and Louisiana. About 125 undisturbed core samples 5 in. in diameter were sliced longitudinally into slabs 1⅝ in. thick and radiographed. Interesting features were selected in 74 slabs. These features were incorporated into test specimens which were 3¼ in. long and 1⅝ in. in square cross section. The specimens were subjected to unconfined compression in a Karol-Warner Model KW500 compression machine driven by a Model KWDV-3 variable speed drive. The deformation versus load was recorded on an automatic strip chart. The rate of strain was selected to provide at least 10 min of loading before failure, that is, the development of the maximum compressive stress. Each test was continued beyond failure to provide sufficient deformation to emphasize the mode of failure. After testing, the specimen was again radiographed. Comparisons were made between pretest and posttest radiography.

Half of all the samples tested showed failure patterns that were determined by features on the pretest radiographs. Some samples, notably those from clay-plug deposits, were lacking in features to begin with so that comparisons were not really possible. Other specimens had been churned so completely by root penetrations or had been desiccated to such a fine, intricate blocky pattern that they behaved as if they were homogeneous materials. Sediments which were layered horizontally and were subjected to compression along a vertical axis did not show failures affected by the bedding. However, tilted bedding planes, and crosslaminated bedding, did produce slippage planes which were controlled by the bedding. These slippages occurred where the bedding was formed by alternating layers of silts or fine sands and clay. Where the material was uniform or contained only subtle evidences of stratification, failure was not affected. The experiments did not provide a usable measure for separating these categories. Secondary mineralizations, though abundant as limonite nodules, did not seem to affect failure patterns, but large, irregular pieces of organic matter did have very distinct effects in determining the positions of failure planes.

The tests, though they indicated that much additional work would be needed in order to relate radiographic features with test behavior, did show that radiography is a promising tool for use in soils testing procedures.

INVESTIGATION OF BASIC STRESS–STRAIN RELATIONSHIPS IN SOILS

From what has been described so far, radiography would appear to be an important, perhaps an indispensible, aid for the investigation of fundamental stress–strain relationships in soils. A number of investigators have been working on the applicability of the method to various aspects of basic soils theory. However, the potential for such studies has hardly been touched and the prospects for future developments in this area are considerable.

Roscoe, Arthur, and James (July, 1963, and August, 1963) went into a detailed discussion of radiography as a method for determining the incremental strains in a soil mass when it is deformed under conditions of plane strain. For plane strain observations, the soil must be contained between two rigid, parallel planes of material which has a low coefficient of friction and is transparent to x rays. Roscoe developed a soil container made from two glass plates 3 ft × 1 ft × 5/8 in. thick and spaced 6 in. apart. The glass plates were held about their periphery by 1½ in. × 1½ in. angle irons bolted to the ends of the container. Radiation was from a 150 kV source with 8 mA rating using a 1.5-mm focal spot. Exposure time was 4 min on Kodak Industrex D film. A network of lead pellets was placed in the soil on a 1-in. grid. Radiographs of boundary strain were taken at regular increments. The displacements of lead shot were determined by measuring the distance between images in two radiographs, compared on an illuminator. Correct alignment was obtained by having a grid of reference lines on the radiographs formed by images of crossed wires glued to the outer face of the glass wall nearest to the cassette.

Observations of pellet displacements were done by two methods. One involved a protractor microscope formed by a Hensoldt Wetzlar microscope,

$20 \times$ magnification, with a linear graduated scale of 0.002-in. divisions in the eye piece, connected to a hollow perspex cylinder which could rotate about the axis of the microscope on a stationary protractor scale. Angles of deviation of the shot pellets were found to be no more than $3°$ from a fixed line, and usually the deviation was much less.

The second method involved placing a radiograph on a light table and holding it with tape. A second radiograph was fixed with tape to a glass plate and the two radiographs were placed in contact. Three raised lugs and a raised strip were affixed to the outer face of the plate. Once the plate was in position, movements were made by micrometer screws pressed against the lugs. The micrometer screws, in turn, were connected to linear inductive transducers. The electrical output from these tranducers was recorded automatically and then presented on punched tape for input to a computer.

Arthur, James, and Roscoe (1964) continued their work with further observations on the determination of stress fields during plane strain of a sand. They noted that if a major discontinuity of strain occurs in the soil mass, then the strains associated with any network of points traversed by this discontinuity are subject to considerable error. However, the presence of major discontinuities could be detected on the radiographs. The authors suggested that such discontinuities do not develop until the peak (failure) loads have been applied.

Using the techniques described above, the authors went on to prepare contour diagrams that illustrated the distribution of shear strain for selected intervals of displacement, maximum values for shear strain, and directions of principal compressive strain. Stresses were derived from the strain relationships and contours were prepared of maximum shear stress and mean stress. Pellet displacements were also plotted on a grid pattern to analyze normal stress distributions on vertical and horizontal planes.

As their studies progressed, Arthur and Roscoe (1965) undertook the examination of edge effects in their models. Particularly, there were questions as to the effects of rigidity, surface finish, and radiation absorption properties of the sides of the soil container. The effects of contact friction between the glass walls and a sand model were investigated by photographing the displacements of a network of nylon hemispheres on the contact face, and comparing them with the movements, detected by x rays, of similar networks of lead shot buried in the sand. Comparisons were made also of side plates of Duralumin, stainless steel, and glass; tests were run on three thicknesses of glass plates: $\frac{1}{4}$ in., $\frac{3}{8}$ in., and $\frac{5}{8}$ in. Glass plates of $\frac{5}{8}$-in. thickness were found to be the most desirable. The experiments showed that side-plate friction was not a large factor in determining the strain and hence the stress field. The rigidity of the container sides affected the forces on the model and also the whole pattern of strain of the sand. It was concluded that the intermediate principal stress plays a significant role in the yielding of the sand and, as a consequence, the Mohr-Coulomb failure criterion is not applicable.

The problems of rupture and changes in void ratios in soils are of great importance in analyzing stress–strain relationships; however, the x-ray and lead-shot techniques described above are not entirely satisfactory in this area. Roscoe (Helenelund *et al.*, 1967) pointed out that the widths of rupture zones are a function of particle size and are independent of the size of the sample. Evidence obtained by x rays for sands and silts and by electron microscopy for clays suggest that rupture widths are on the order of 10 to 20 particle "diameters" or "thicknesses." The problems involved in studying soils under postpeak stress-ratio conditions increase greatly as grain sizes diminish. For coarser materials, pore volume changes may precede rupture and there are grain slips which accompany shear distortion.

Roscoe reported that Coumoulos (1967) had tested a gamma-ray device which supplemented x radiography in this area. Coumoulos used a 300-millicurie americium-241 isotope with a 458-year half life. The beam was collimated to a $\frac{3}{16}$-in. diameter and transmitted to the sample with an energy level of 60 kV. Rays passing through the sample were detected by a sodium iodide crystal connected to a photomultiplier. The source and the detector were on a common mounting so that the beam could be traversed across any part of the specimen. The recordings were calibrated at 10,000, for a dense sand, and 20,000, for a loose sand. Sample thickness was 10 cm. It was found that the local voids ratio at any point in these samples could be determined with an error of less than 0.5%.

It may be noted that Sopp (1964) first observed that x radiography was a useful tool for determining shear planes and shear angles in laboratory testing. Leitch and Yong (1967) used density analyses of x radiographs, combined with optical observations, to suggest that shearing of a normally consolidated clay is accompanied by a reorientation of the mineral particles so that the long axes parallel the shear plane in its immediate vicinity and there is an appreciable, though decreasing, reorientation away from the plane.

Kitani and Persson (1967) used radiography to observe stress–strain relationships in a cylindrical soil sample which was compressed along its longitudinal axis and was laterally confined with a flexible sidewall, held together by springs of various spring rates. The sample was a remolded loam, 3 in. in diameter and 7 in. high. Lead spheres were imbedded in the soil. Shear strain was recorded on the x-ray picture with exposures every 2.5 min and the radiography was done without halting the compression. The data permitted the establishment of relationships between stress and strain in the specimens, both longitudinally and laterally, and related the influence of spring rate on relationships between shear stress and shear strain. The effect of wall friction was found to be negligible as there was a good agreement between shear strain obtained from x radiographs and values calculated from measured principle strains.

In conventional triaxial tests, internal strains that have been evaluated were axial strains. Most electrical displacement devices or strain measuring devices placed in triaxial specimens are bulky, limited in their range of measurements, and likely to be a source of disturbance in the testing. In order to avoid these handicaps, Kirkpatrick and Belshaw (1968) undertook to apply radiographic examinations in triaxial testing.

Figure 9-11 shows the apparatus used in their testing. The outer cell and

Fig. 9-11. Apparatus used in radiographic examination of triaxial test. Note absence of restraining fluid and the shaped sandbag behind the specimen for compensation of the cylindrical shape. (Courtesy W. M. Kirkpatrick.)

fluid which normally would apply the radial stress to the samples has been dispensed with as these would reduce the penetration of x rays and would cause radiation scatter. A radial stress was applied by encasing the specimens in a membrane and evacuating the inside. In order to obtain evenly exposed radiographs, the cylindrical shape of the specimens was compensated for by a shaped sandbag placed behind the specimens. Each specimen was made of remolded sand and was 10 in. in diameter and 20 in. high. Ninety-nine pieces of lead shot were placed in each specimen in nine rows and eleven columns on an approximately 1-in.-square grid in a vertical plane through the center of the specimen. The focal distance from x-ray source to lead shot was approximately seven times the distance of the lead shot to photographic plate. Alignment screens made of perspex plates with fine steel wires inlaid into the surface were placed before and behind the specimen. The x-ray film was clamped onto the rear screen.

Radiographs were taken prior to testing and at four different axial strains in each test. After each exposure, the sample was unloaded, rotated 30° and another exposure was made. The sample was then rotated to its original position and the loading was continued.

Testing included comparisons between the effects of rough platens and smooth, lubricated platens.

Strain measurement was done by comparing radiographs. Coordinate positions of shot displacements were recorded on punched tape using an electronic x, y coordinate table. Accuracy was believed to be to 0.1 mm. The analysis was programmed to a computer. Information on shot displacements from a selected radial plane, and the axial and shear strains were calculated. Curves were developed which showed the axis of radial strain from the geometric axis at various axial strains. The authors concluded that the x-ray technique is reasonably successful for measuring internal displacements in the axially symmetrical problem of the compression cylinder.

The progress of consolidation in organic soils is a special problem in which radiography was found to be useful. Wilson and Lo (1966) prepared samples of amorphous, granular peat from Parry Sound, Canada. Water content at the start of the test was 725 % of dry weight. A lucite consolidometer, 4½ in. in diameter by 6 in. high, was made. The piston was fitted with a porous stone and loads were applied by deadweight. Measurement of pore water pressure was made through an entrant in the base of the consolidometer. The consolidometer was slowly filled with peat. Three-inch lengths of wire solder, 0.015 in. thick, were placed horizontally through the centerline at ½-in. intervals vertically. Horizontal and vertical markers were used as reference lines outside the consolidometer.

Radiography was done with a 130 kV apparatus with a focal distance of 30 in. Kodak x-ray film, type KK, was used with 0.005-in. lead intensifying screens. The x-ray unit was left running continuously but exposure of the x-ray film was done periodically by opening a lead shutter.

The tests indicated that the upper layer consolidates rapidly but underlying layers take longer, possibly because of a rapid change in permeability in the upper layer. Consolidation at the base of the sample did not take place until consolidation was nearly complete in the surface layer.

In agriculture, problems of poor drainage, unexpectedly low crop yield, and cloddy seed beds may be the result of soil compaction from transient loading, as from vehicles. Chancellor, Schmidt, and Soehne (1962) have applied x radiography to these problems. Preliminary tests were made with a piston loaded onto a remolded soil marked with lead pellets. In addition to tracing pellet displacements, these authors suggested that a qualitative evaluation of density changes could be made from the tonal quality of the radiographic film. Shear planes were identified on the radiographs and contours were made on radial sections showing lines of equal volume reduction.

Further work by Chancellor and Schmidt (1962) explored the effects of wheel movement on soil movement, soil density increase, and dimensional soil deformation. Tests were on clay loam and sandy loam. Radiographic techniques were the same as in the previous tests. It was concluded that for agricultural soils a large portion of the volume of surface impression is absorbed as compaction within the soil mass. Compaction was found to be accompanied by shearing deformation in most cases.

The response behavior of sand soils under moving wheels was studied by Yong, Boyd, and Webb (1967) and response behavior of clay soils was reported on by Yong, Fitzpatrick-Nash, and Webb (1968). Their facility involved a flash x-ray unit capable of pulsing up to 10 exposures, at 300 kV, at a maximum rate of 2 pulses per sec. Their model included a test track with a total length of 8 ft and a depth varying between 6 in. and 2 ft. The container was 8 in. wide, made of 1-in. plywood and sealed on the inside with waterproof paint. The interior sidewalls were $\frac{1}{4}$ in. glass, strengthened by $\frac{1}{4}$-in. plywood backing and smeared with Vaseline to reduce wall friction.

The source-to-film distance was 16.5 in. so that a 7×7 in. area was viewed in the plane of the wheel centerline. The tracer objects imbedded in the soil were specially designed, three-dimensional, orthogonal crosses, 7 mm in overall dimension, made from 94% lead and 6% antimony. Kodak Royal Blue Medical film was used along with Dupont calcium tungstate intensifying screens. The cassette holder contained four lead markers on the side receiving the radiation. A fifth lead marker was placed in a position corresponding to the optical center.

The x-ray pulses were controlled by microswitches placed so that exposures were made when the wheel was 6.1 in. before, exactly over, and 6.1 in. beyond the centerline of the x-ray source. A moving cassette holder was

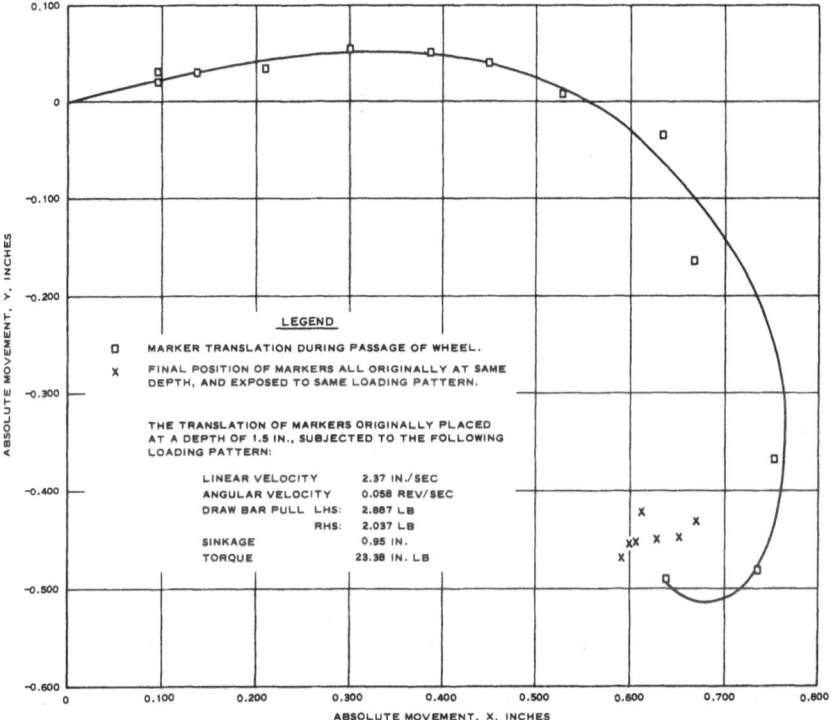

Fig. 9-12. Translation of markers by a moving wheel from determinations by flash radiography (after Yong, Boyd, and Webb, 1967).

designed in order to move freshly loaded x-ray film into position for these exposures.

Analysis of the radiographic images first involved transference of the marker points onto a translucent "Matex" acetate sheet using the reference grid to align the radiographs. The points were then transferred to graph paper and registered with an x, y plotter onto a printer by a Dymec 2010B Automatic Scanning Digital Voltmeter. The information was then punched manually onto IBM cards. The image coordinate system was now corrected for optical errors using an IBM 7044 Digital Computer. This correction placed the points into the centerline plane of the wheel trace.

An example of translation patterns in markers placed at a depth of 1.5 in. and for a specified condition of loading is shown in Fig. 9-12. The time factors in the radiographic exposures permitted analyses to be made of x and y components of velocity history, particle paths and particle acceleration, and torque versus slip rate. Marker displacements permitted calculation of inferred density changes in the soil mass.

BIBLIOGRAPHY

Arthur, J. R. F. and K. H. Roscoe (1965). An examination of the edge effects in plane-strain model earth pressure tests, *Proc. Sixth Int. Conf. on Soil Mech. & Found. Eng.*, *Montreal*, Univ. Toronto Press, Vol. 2, 5/1, pp. 363–367.

—— R. G. James, and K. H. Roscoe (1964). The determination of stress fields during plane strain of a sand mass, *Geotechnique* **14**: 283–308.

Baker, W. J. and Frank J. Janza (1966). *Investigation of Flash X-Ray Techniques in Soil Dynamics and Interaction Problems*, TR No. AFWL-TR-66-50, Air Force Weapons Laboratory, Kirtland AFB, N. Mex., 58 pp.

Berdan, D. and R. K. Bernhard (1950). Pilot studies of soil density measurements by means of x-rays, *Proc. Am. Soc. Test Mat.* **50**: 1328–1339.

Bergfelt, Allan (1956). Loading tests on clay, *Geotechnique* **6**: 15–31.

Bloedow, F. H. (1962). *Radiographic Instrumentation Study*, AFSWC-TDR-62-44, Kirtland AFB, N. Mex., 143 pp.

Chancellor, W. J. and R. H. Schmidt (1962). Soil deformation beneath surface loads, *Trans. ASAE* **5** (2): 240–247.

——, ——, and W. H. Soehne (1962). Laboratory measurement of soil compaction and plastic flow, *Trans. ASAE* **5** (2): 235–239.

Clark, G. L. (1955). *Applied X-Rays*, fourth edition, McGraw-Hill, New York, 843 pp.

Colp, John L. (1968). *Terradynamics: A Study of Projectile Penetration of Natural Earth Materials*, Sandia Laboratories SC-DR-68-215, Albuquerque, N. Mex., 61 pp.

Coumoulos, D. G. (1967). *A Radiographic Study of Soils*, doctoral dissertation, Cambridge University.

Davis, H. E. and R. J. Woodward (1949). Some laboratory studies of factors pertaining to the bearing capacity of soils, *Hwy. Res. Board, Proc. 29th Ann. Meeting, Washington*, pp. 467–476.

Gerber, Emil (1929). *Untersuchugen über die Druckverteilung im örtlich belasteten Sand*, doctoral dissertation, Zurich, Switzerland.

Gornak, George (1967). *Radiographic Study of Ground Shock and Gas Cavity Expansion*, report 1896, U. S. Army Eng. Res. & Dev. Lab., Fort Belvoir, Va., 37 pp.

Helenelund, K. V. *et al.* (1967). Shear strength of soil other than clay, *Proc. Geotech. Conf. on Shear Strength Properties of Natural Soils & Rocks*, Vol. 2, Norwegian Geotech. Inst., Oslo, pp. 187–221.

Kirkpatrick, W. M. and D. J. Belshaw (1968). On the interpretation of the triaxial test, *Geotechnique* **18** (3): 336–350.

Kitani, O. and Sverker P. E. Persson (1967). Stress–strain relationships for soil with variable lateral strain, *Trans. ASCE:* pp. 738–745.

Krinitzsky, E. L. (1970). *Radiographic Study of the Effects of Geological Features on the Strength Properties of Undisturbed Soils*, miscellaneous paper, Waterways Experiment Station, Vicksburg, Miss. (in press).

Leitch, H. C. and R. N. Yong (1967). The rate dependent mechanism of shear failure in clay soils, *Soil Mech. Ser. No. 21*, McGill Univ., Montreal, 140 pp.

McLaren, A. D., R. A. Luse, and J. J. Skujins (1962). Sterilization of soil by irradiation and some further observations on soil enzyme activity, *Proc. Soil Sci. Soc. of Am.* **26** (4): 371–377.

O'Dea, James J., Jr. (1949). *An Experimental Investigation of the Vibratory Characteristics of Sand*, master's thesis, Rensselaer Poly. Inst., Troy, N. Y., 102 pp.

Perry, E. B. (1968a). *The Effect of Shallow Burial on the Load-Displacement-Time Response of Square Footings in Clay under Impulsive Loading*, master's thesis, Mississippi State University, State College, Miss., 121 pp.

—— (1968b). *Application of Radiography to the Study of Small-Scale Footing Tests in Clay*, M. P. S-68-12, Waterways Experiment Station, Vicksburg, Miss., 15 pp.

Robinsky, E. I. (1963). *Effect of Shape and Volume Displacement on the Capacity of Piles in Sand*, doctoral dissertation, University of Toronto, Toronto, 60 pp.

—— and C. F. Morrison (1964). Sand displacement and compaction around model friction piles, *Canadian Geotechnical Journal* 1 (2): 81–93.

Roscoe, K. H., J. R. F. Arthur, and R. G. James (July, 1963, and August, 1963). The determination of strains in soils by an x-ray method, *Engineering and Public Works Review* **58**: 873–876, 1009–1012.

Sopp, O. I. (1964). X-ray radiography and soil mechanics: localization of shear planes in soil samples, *Nature* **202**: 832.

Tschebotarioff, Gregory P. (1950). Discussion to paper by D. Berdan and R. K. Bernhard: pilot studies of soil density measurements by means of x-rays, *Proc. Am. Soc. Test Mat.* **50**: 1338–1339.

Watkins, R. K. (1957). *Characteristics of the Modulus of Passive Resistance of Soil*, doctoral dissertation, Iowa State College, Ames, Iowa, 182 pp.

Wilson, Nyal E. and M. B. Lo (1966). The progress of consolidation in an inorganic soil, *Proc. Eleventh Muskeg Res. Conf.*, Tech. Memo 87, Nat. Res. Counc., Ottawa, I. 1, pp. 1–12.

Yong, R. N., C. W. Boyd, and G. L. Webb (1967). Experimental study of behaviour of sand under moving rigid wheels, *Soil Mech. Ser. No. 20*, McGill Univ., Montreal, 79 pp.

—— J. D. Fitzpatrick-Nash, and G. L. Webb (1968). Response behaviour of clay soil under a moving rigid wheel, *Soil Mech. Ser. No. 23*, McGill Univ., Montreal, 71 pp.

—— R. D. Japp, and H. C. Leitch (1964). Strain rate effect on clay soil strength, *Soil Mech. Ser. No. 12*, McGill Univ., Montreal, 161 pp.

Chapter 10

MICRORADIOGRAPHY

There are no successful x-ray optical instruments that give enlarged true images of an object, but an effect of optical enlargement is achieved by two processes known as contact microradiography and x-ray projection microscopy.

A third method, known as x-ray reflection microscopy (Guinier and Dexter, 1963; Kirkpatrick, 1967; Clark, 1963), makes use of optical principles, but has never been fully developed. X-ray reflection microscopy involves registering x-rays that are reflected at grazing incidence with a mirror. A smooth, solid surface can refleet x-rays efficiently if the beam meets the surface at a small enough angle: 1-Å wavelength reaching a glass surface must meet at an angle of about 1° to to be reflected satisfactorily. An instrument using this principle has been constructed on a research and development basis (McGee, 1957), however, there are problems which are as yet unresolved. Low-angle incidence produces astigmatic effects and image aberrations, diminished field, and low resolution.

CONTACT MICRORADIOGRAPHY

The contact process (Engström, 1967; Bertin and Samber, 1967; Clark, 1963) involves registration from specimen to film without enlargement. The film in this case must contain a fine-grained emulsion so that it can be considerably enlarged or examined with an optical microscope. The specimen of rock or soil must be in a very thin slice, preferably on the order of $10\,\mu$ to $35\,\mu$. With thin specimens, x-ray absorption is weak and good contrast can be produced only by very long wavelengths; consequently, low voltages are used. Soft radiation has the further advantage that photoelectrons are not liberated in the specimen or in the photographic emulsion and, thus, scatter effects are avoided and a sharper image is obtained. The work must be done in a vacuum as the low-energy radiation may be absorbed within a few centimeters of air.

Wavelengths are classified in the following manner:

135

Ultrahard: <0.1 Å
Hard: 0.1 to 1.0 Å
Soft: 1.0 to 10.0 Å
Ultrasoft: >10.0 Å

Selections of wavelengths can be made by using a selection of targets. Sample values that are obtainable are 8.34 Å for Al, 13.3 Å for Cu, 21.7 Å for Cr, and 27.4 Å for Ti. The voltage must be kept low in order to excite only the $L\alpha$ portion of the characteristic spectrum, and filters can be used to narrow the range of wavelengths in the spectrum.

An apparatus used for contact microradiography is shown in Fig. 10-1. The x-ray tube and the camera are constructed together. The focal spot on the target is on the order of 0.1 mm. A filter prevents light generated by the apparatus from entering the camera chamber. The specimen and the film are placed in contact and the apparatus is evacuated to a vacuum. The apparatus operates at about 1.5 kV and 6 mA with a focal distance of 5 cm or less.

The degree of resolution is potentially the same as that for optical

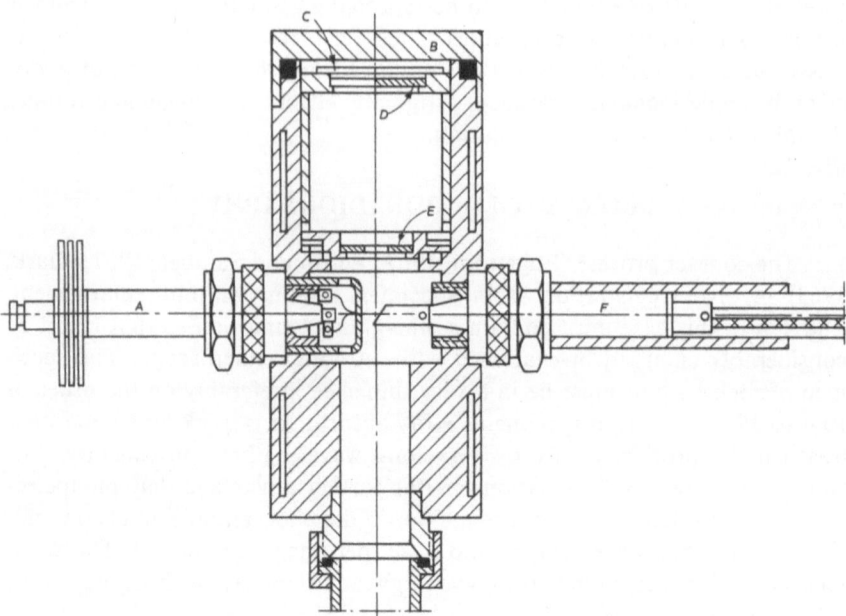

Fig. 10-1. Apparatus designed for contact microradiography (after Engström). A, cathode; B, removable lid for loading camera; C, photographic film; D, specimen; E, light filter; F, anode.

microscopy. The intensity recorded at points on the film is a function of the mass absorption coefficient and the mass per unit area of the sample. Hence, quantitative evaluations may be made.

The method has been used for microradiography more widely than any other (see Clark, 1963, p. 610). It was applied as early as 1913 by Goby in investigations of the internal structure of diatoms and foraminifera, and in 1925 he produced stereoscopic microradiographs. Since then there have been many varied applications in the biological sciences and in metallurgy. Cohen and Schloegl (1960) showed the method to be an improvement over thin-section petrographic microscopy and reflection microscopy when small particles of lead minerals, anglesite and cerussite, must be distinguished from gangue minerals. The method also helps to explore the occurrence of gold particles, a few microns in diameter, inside crystals of pyrite or other sulfides, in evaluating the distribution of iron in iron ores, sinters, and slags.

ELECTRON MICRORADIOGRAPHY

A special form of contact microradiography is one that registers on photographic film the secondary photoelectrons which are released by materials when they are bombarded by x rays (Trillat, 1959; McGonnagle, 1961; Clark, 1963). This technique is generally known as electron microradiography but may also be referred to more simply as electron radiography.

Schematic diagrams of electron microradiography are shown in Fig. 10-2 for electron reflection (A) and electron penetration (B). The specimen may be in the form of a petrographic thin section, for penetration by electrons, or either a thin section or small slice polished on one side, for reflection of electrons.

Film used is of fine-grained, thin emulsion types that permit passage of x rays without absorption but which register electron activity. Ultrahigh resolution spectroscopic plates are satisfactory for this work. The emulsion must be placed as close to the specimen as possible, and the specimen must be cut smoothly or its topography will be registered. X rays are in the form of hard radiation produced by voltages between 100 and 200 kV or higher. Soft radiation must be filtered out as it contributes to film clouding and produces negligible photoemission. Lead is the most satisfactory material for the electron-producing foil because of its high atomic number and high electron density. Foil thickness may be on the order of 25 μ and can be pressed against the specimen. As the electrons are emitted by the foil in all directions and with varying energies there is a reduction in sharpness of image.

The reflection technique registers differentials in the emission of electrons from the surface of the specimen and the penetration method registers the passage of electrons from the lead foil through the specimen and onto the

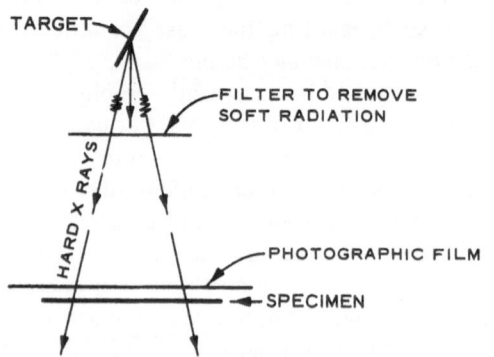

a

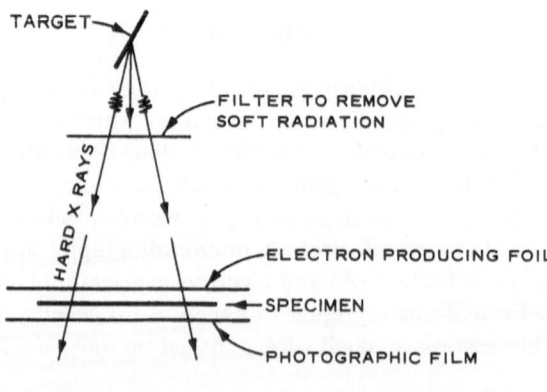

b

Fig. 10-2. Electron microradiography by reflection (a) and by penetration (b).

photographic emulsion. It may be noticed that electron microradiography differs from other forms of microradiography in two respects:

1. High powered x-ray equipment is used.
2. With the electron-reflection technique, the examination is of the surface of a specimen rather than of superpositioned internal structures.

Trillat (1959) describes work he has done with the method in studying minerals. The reflection technique appears to work well with opaque, heavy metallic minerals and with radioactive minerals. With radioactive minerals, the technique may be used in combination with autoradiation in which inherent radiation properties of the mineral are registered. In general, however, the method has less promise for earth studies than other forms of microradiography.

X-RAY PROJECTION MICROSCOPY

The x-ray projection microscope (Ong, 1959, 1961, and 1967; Guinier and Dexter, 1963; Clark, 1963) is essentially an x-ray tube in which the electron beam is brought to a very tiny focus on the target (Fig. 10-3). Generally, the focal spot used has a diameter of 0.1 to 1.0 μ. An x-ray emission from a source of this dimension is not too feeble to provide registration, it does not induce scatter, and it is precise enough to furnish an image resolving power equal to that of an optical microscope. The target is in the form of a thin, replaceable foil which is both an anode and a window. The metal in the target can be selected, as in contact microradiography, to provide desired wavelengths of soft to ultrasoft radiation. Multiple exposures of the same specimen with differing radiation wavelengths allow the study of absorption characteristics and are useful in interpreting the properties of the specimen. Absorption coefficients can be calculated from tables that are available both for the common elements and for the radiation wavelengths used.

The sample can be placed very close to the source of radiation, with distance from the focal spot to the specimen on the order of 0.2 mm. At a film distance of 2 cm, there is an enlargement factor of 100. The film used is of fine-grained emulsion types and is susceptible to further enlargement either by photoprinting or viewing with an optical microscope. Imaging may be done on a fluorescent frame prior to registering on film.

As with most other radiography, there is no limiting depth of field and all levels of the sample are in focus. Thus, stereographic views are possible.

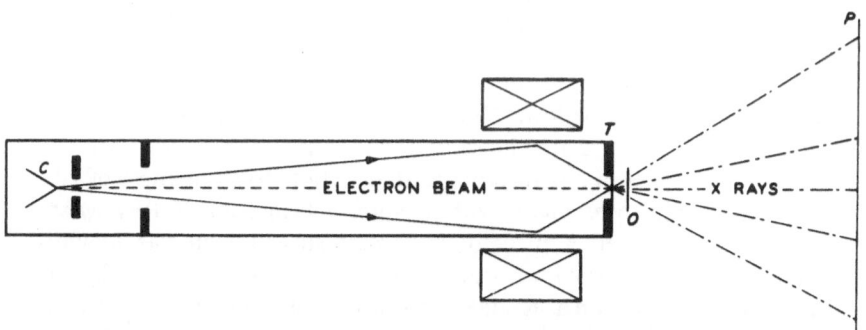

Fig. 10-3. Schematic design for an x-ray projection microscope (after Cosslett). C, cathode; T, target; O, object; P, photographic film.

The apparatus is available as an accessory to certain electron micro-scopes.

The method lends itself, for example, to determinative analyses of cementing material in sandstone (Ong, 1967), and to analyses of the spatial distribution of iron components in an olivine gabbro (Ong, 1961). However, despite these demonstrations, there has been as yet no significant utilization of this method in earth studies, partly because optical methods, using the petrographic microscope, are much more versatile and more readily available to geologists and soils engineers. Microradiography does have some poten-tialities for soils studies which may advance its applications in the future.

The strength values and the processes of deformation under loading in clays and clay shales are largely dependent on a combination of clay mineral-ogy, the fabric of clays or the clay particle structure, and the changes that are induced by absorption or release of ground moisture and by cation ex-changes with pore fluids. These aspects are contributing factors and must be evaluated in basic research on such problems as liquefaction in clays, ex-pansion or "heaving" in clays, and changes in shear strength in clays and clay shales.

Improvements in the techniques for the study of fabrics in clays and pore structures in clays are needed. The earlier concept of house-of-cards struc-ture versus plane-parallel structure (Mitchell, 1956) is now regarded as oversimplified. Clays are sedimented in nature as floccules or aggregates of clay particles that are much larger than the size ranges of the component clay minerals. Aggregates of these floccules (or floc units according to Yong and Warkentin, 1966) may differ significantly in their fabric and in their susceptibility to deformation. With loading, their patterns may be altered in significantly different ways.

Research in these areas must make use of many techniques including petrographic microscopy, x-ray diffraction, and scanning electron micros-copy. Microradiography has a place in this work as it can add certain facets of information that are not as readily available by the other methods. Figure 10-4 shows a microradiograph of undisturbed, naturally sedimented clay from a channel fill deposit in the lower Mississippi Valley. The depth from which the sample was taken is 72 ft. The sample was impregnated with Carbowax and cut, in the manner of a petrographic thin section, to a thickness of $20\,\mu$. The microradiograph was made with a Norelco Projection X-Ray Micro-scope using an Al target. A piece of 1500-mesh silver screen was mounted next to the specimen to provide a scale. Exposure was for 1 min at 15 kV and $40\,\mu$A. Registration was on a Kodak lantern slide plate and printing was on No. F-5 paper. The same specimen was also photographed using Ti, Cu, Fe, Ni, Ag, and Cr targets. The sharpest images were produced by the Al, Ti, and Ag targets. Note that the floccule structure of the clay can be examined. The

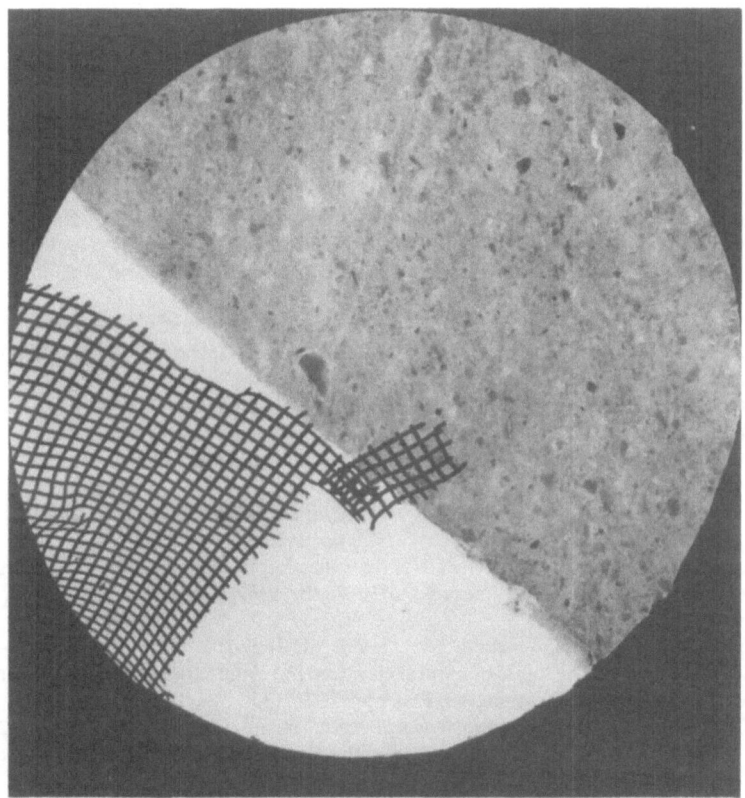

Fig. 10-4. Microradiograph of clay from a channel fill deposit in the lower Mississippi Valley made with a projection x-ray microscope; depth 72 ft. Thickness of slice, 20 μ. 1500-mesh silver screen. Al target.

dark minerals are mostly silt size particles of quartz. The same slide under a petrographic microscope would fail to show the details of clay structure even at the same order of magnification. X rays in these circumstances tend to reveal finer detail than is possible with visible light. Resolving power is affected both by wavelength of rays and differential absorption in the material. Soft x rays provide finer resolutions as their wavelengths are 1 to 10 Å, whereas light passing through optical microscopes ranges from 4000 to 7000 Å. The slide thickness of 20 μ is thinner than the standard thickness of 35 μ used for optical petrography, thinner slices being necessary to avoid excessive superposition of floccule outlines. Recent work, notably by Roland Pusch (1964), shows that even thinner slices of undisturbed clays can be made by slicing quick-frozen specimens with a microtome having a freezing chamber;

slices 8 to 15 μ in thickness can be produced with a minimum of effort. It is practicable to examine microtome slices both by optical petrographic means and by microradiography. X-ray projection microscopy holds considerable promise of future applicability for this phase of earth studies.

BIBLIOGRAPHY

Bertin, E. P. and R. L. Samber (1967). Contact microradiography, *in:* E. F. Kaelble (ed), *Handbook of X-Rays*, McGraw-Hill, New York, Chapter 45, pp. 3–28.

Clark, George L. (ed) (1963). *The Encyclopedia of X-Rays and Gamma Rays*, Reinhold, New York, pp. 604–621.

Cohen, E. and I. Schloegl (1960). The application of microradiography in mineral dressing, *in:* Arne Engström, A. Y. Cosslett, and H. Pattee (eds), *X-Ray Microscopy and X-Ray Microanalysis*, Elsevier, London, pp. 133–148.

Engström, Arne (1967). Contact microradiography with ultrasoft x-rays, *in:* E. F. Kaelble (ed), *Handbook of X-Rays*, McGraw-Hill, New York, Chapter 46, pp. 1–10.

Guinier, A. and D. L. Dexter (1963). *X-Ray Studies of Materials*, Interscience Publishers, New York, pp. 17–22.

Kirkpatrick, Paul (1967). Reflection microscopy, *in:* E. F. Kaelble (ed), *Handbook of X-Rays*, McGraw-Hill, New York, Chapter 48, pp. 1–10.

McGee, J. F. (1957). A long-wavelength x-ray reflection microscope, *in:* V. E. Cosslett, Arne Engström, and H. H. Pattee (eds), *Microradiography*, Academic Press, New York, p. 164.

McGonnagle, W. J. (1961). *Nondestructive Testing*, McGraw-Hill, New York, pp. 168–170.

Mitchell, J. K. (1956). The fabric of natural clays and its relation to engineering properties, *Proc. Hwy. Res. Board*, Washington, D. C. pp. 693–713.

Ong, Poen Sing (1959). *Microprojection with X Rays*, Martinus Nijhoff, The Hague, 132 pp.

―――― (1961). Isolation of selected elements with an x-ray projection microscope, *Norelco Reporter*, **8** (1): 7.

―――― (1967). X-ray projection microscopy, *in:* E. F. Kaelble (ed), *Handbook of X-Rays*, McGraw-Hill, New York, Chapter 47, pp. 1–24.

Pusch, Roland (1964). *On the Structure of Clay Sediments*, Handlingar No. 48, Byggforskningen, Stockholm, 60 pp.

Trillat, Jean-Jacques (1959). *Exploring the Structure of Matter*, transl. by F. W. Kent, George Allen & Unwin, London, pp. 30–47.

Yong, R. N. and B. P. Warkentin (1966). *Introduction to Soil Behavior*, Macmillan, New York, pp. 108–111.

Chapter 11

IMAGE QUANTIFYING

Measurements of film density on a radiographic image lend themselves to interpretations of soil density and permit the analysis of the patterns of internal density adjustments which accompany deformation. The material tested, however, must be homogeneous and must be of a known composition, and comparative radiography must be done on equal thicknesses of sample.

ISODENSITY TRACING

Isodensity tracing is the photometric measurement of film density in a radiograph. Photodensitometer devices are available which measure points. Other units measure linear traces and record them as graphs, and still more sophisticated units are able to scan a radiograph with a multitude of parallel traces which are printed with varying modifications of line intensity, showing up as shades of gray in a pictographic printout. Other tracing units are color coded and are able to print film density changes in patterns of up to four colors plus black. It is possible also to put density readings onto magnetic tape if subsequent processing by digital computer is required.

An example of image quantizing with radiographic film density values is shown in Fig. 11-1. The sample is a nearly featureless prodelta clay from a depth of 160 ft in the deltaic plain of the Mississippi River. Previous stress history of the clay has caused it to deform very much as a brittle material. Electronic heightening of the film density differentials causes a striking printout of what is really an isodensity pattern of soil density variations. Control values for densities may be obtained either by direct measurement on portions of the soil sample or by interpretative methods to be discussed later in this chapter. Isodensity contours may then be drawn if they are desired.

Another example of a heightening of density values is shown in Fig. 11-2. The sample is a remolded clay which has been subjected to dynamic plate loading. A radial section, $3/8$ in. thick has been cut out from under the plate. Note that the increased densities beneath the plate can almost be

ST. BERNARD PRODELTA CLAY (−160.0 FT)

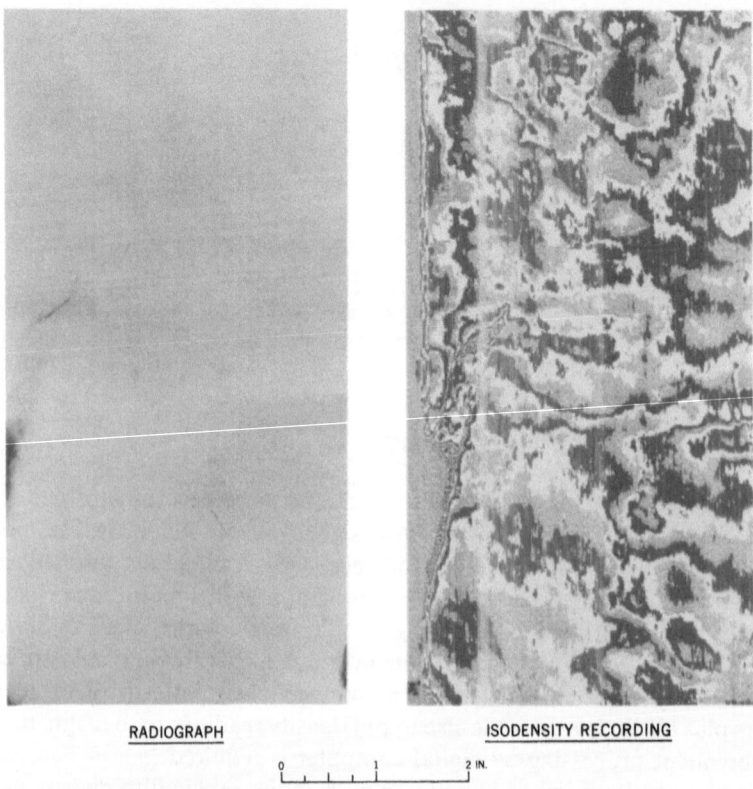

RADIOGRAPH   ISODENSITY RECORDING

0   1   2 IN.

Fig. 11-1. Isodensity recording from radiograph of a nearly featureless prodelta clay. Sample from along the Mississippi River near Head of Passes, depth 160 ft. Recording by Technical Operations, Inc.

contoured. Yet, in the radiograph, the density changes are so faint that they can hardly be recognized.

Figures 11-3 and 11-4 show a photograph, a radiograph, and varied quantized printouts made from a complicated brecciated zone within the clay shale foundation at Warm Springs Dam in California. Note that fractures show up best in the radiograph and the radiograph displays the relation between breccia fragments and the matrix. The quantized printouts in Fig. 11-4 were selected with differing degrees of detail based on selection of film isodensity intervals. Hence, there is a latitude in selecting and forming the printouts which show optimum density relationships.

The differential tracing (Fig 11-4) is not an isodensity print but a measure

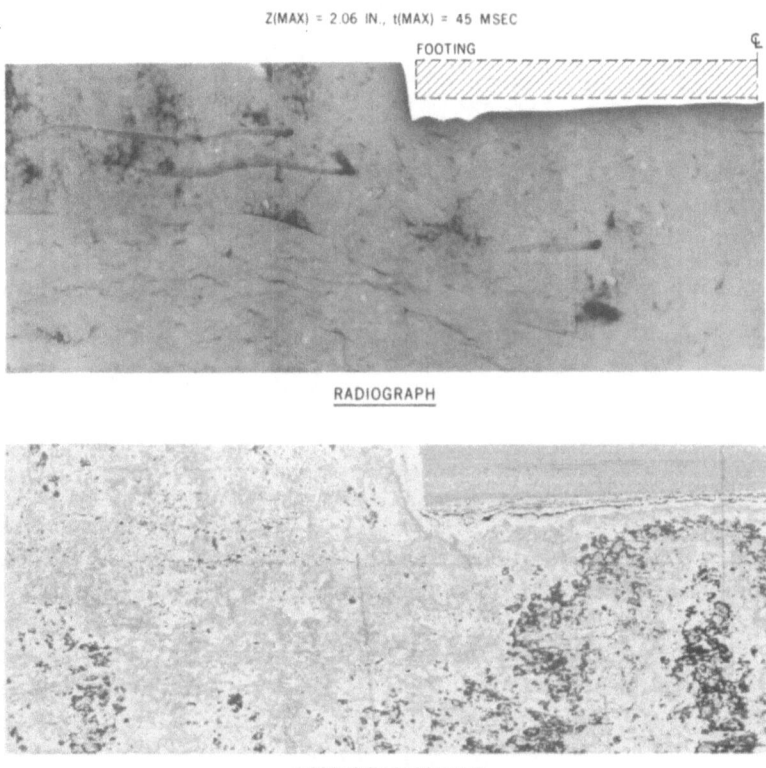

DYNAMIC TEST 58–1 (9.06 KIPS)

Z(MAX) = 2.06 IN., t(MAX) = 45 MSEC

FOOTING

RADIOGRAPH

ISODENSITY RECORDING

Fig. 11-2. Isodensity recording from radiograph of a remolded clay subjected to dynamic plate loading. Recording by Technical Operations, Inc.

of interruptions in the orderly progression of film density changes. Continuous progressions are printed as gray, an interruption in the progression as black. The technique has supplementary usage in picking up shear zones or fracture zones which may have been obscured in the other views.

CALCULATION OF RADIATION ABSORPTION

The absorption of x rays or gamma rays passing through a soil may be calculated by the exponential decay equation

$$I = I_0 e^{-\mu x}$$

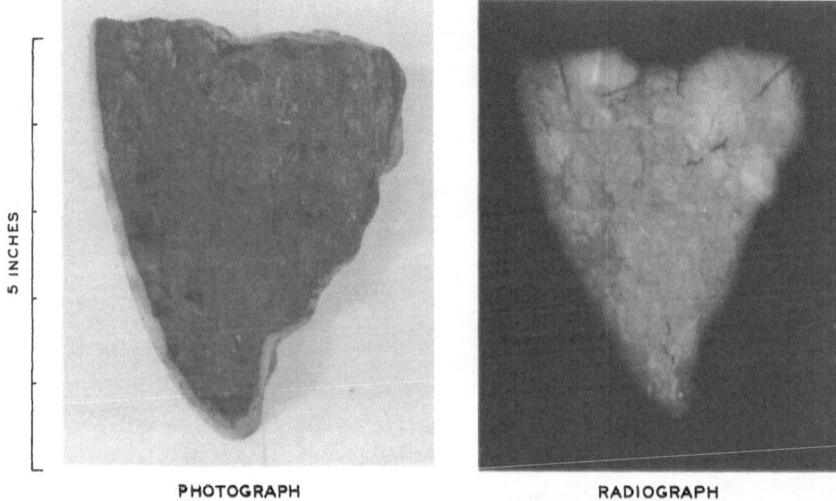

PHOTOGRAPH   RADIOGRAPH

Fig. 11-3. Photograph and radiograph of $\frac{3}{8}$ in. slice of brecciated clay shale from foundation of Warm Springs Dam, California.

where I is the intensity of transmitted radiation, I_0 is the intensity of incident radiation, μ is the absorption coefficient, and x is the path length through soil.

The absorption coefficient for a soil must be expressed in terms of the density of the mineral constituents and the density of the aggregate mass. Thus,

$$I = I_0 e^{-(\mu/\rho)\rho_T x}$$

where μ/ρ is the mass absorption coefficient for the minerals composing a soil and ρ_T is the density of the soil mass. The coefficients for mass absorption of radiation vary for different levels of radiation energy. Figure 11-5 shows the variations in coefficients for a group of the common elements according to changes in radiation energy. These graphs were developed by Bloedow (1962) from tables published by McMaster (1959). McMaster itemizes the individual effects of scattering, photoelectric changes, and pair formation together with total mass absorption and linear absorption for energies from 0.01 to 30.0 MeV. These tables are presented for all of the common elements.

Calculation of the mass absorption coefficient μ/ρ for a compound from tables which show only the values for elements may be done by summation:

$$\left(\frac{\mu}{\rho}\right)_{compound} = R_1\left(\frac{\mu}{\rho}\right)_1 + R_2\left(\frac{\mu}{\rho}\right)_2 + \cdots + R_n\left(\frac{\mu}{\rho}\right)_n$$

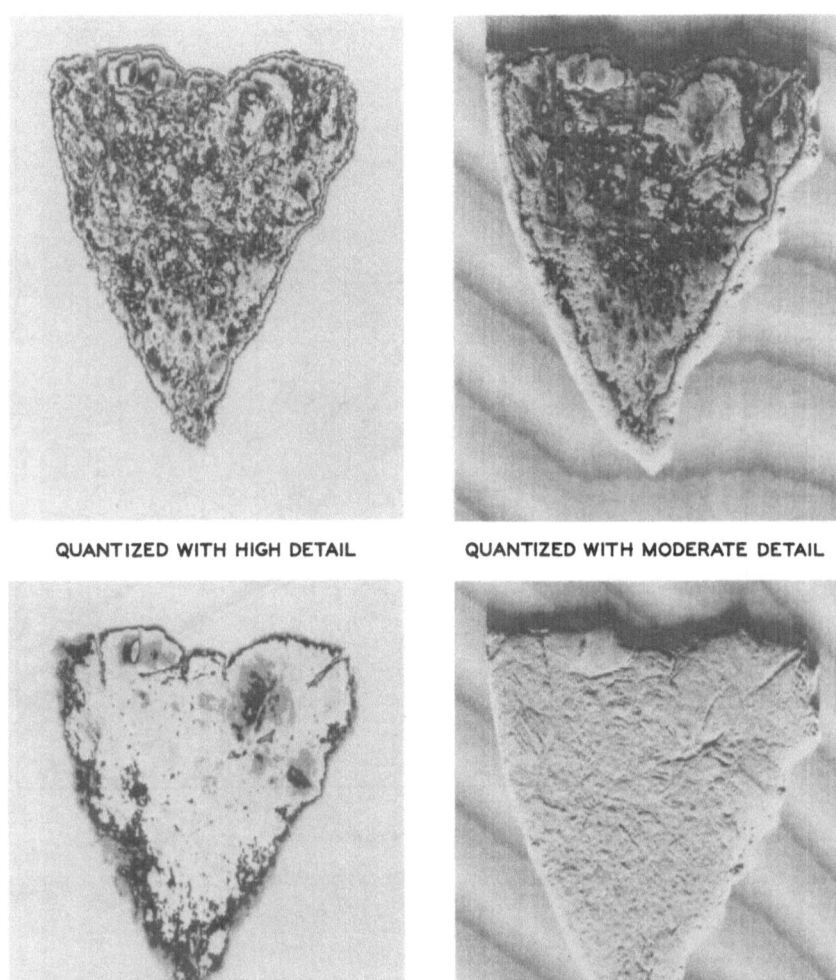

QUANTIZED WITH HIGH DETAIL   QUANTIZED WITH MODERATE DETAIL

QUANTIZED WITH LOW DETAIL   DIFFERENTIAL TRACING

Fig. 11-4.   Quantized prints, with varying detail, and a differential tracing of the sample in Fig. 11-3. Recording by Technical Operations, Inc.

where $(\mu/\rho)_1$ is the mass absorption coefficient of element 1 for a given energy of radiation and R_1 is the ratio of the atomic weight of element 1 in the compound to the total atomic weight of the compound. Bloedow furnished an example of the calculation for kaolinite $(2H_2O \cdot Al_2O_3 \cdot SiO_2)$ radiographed

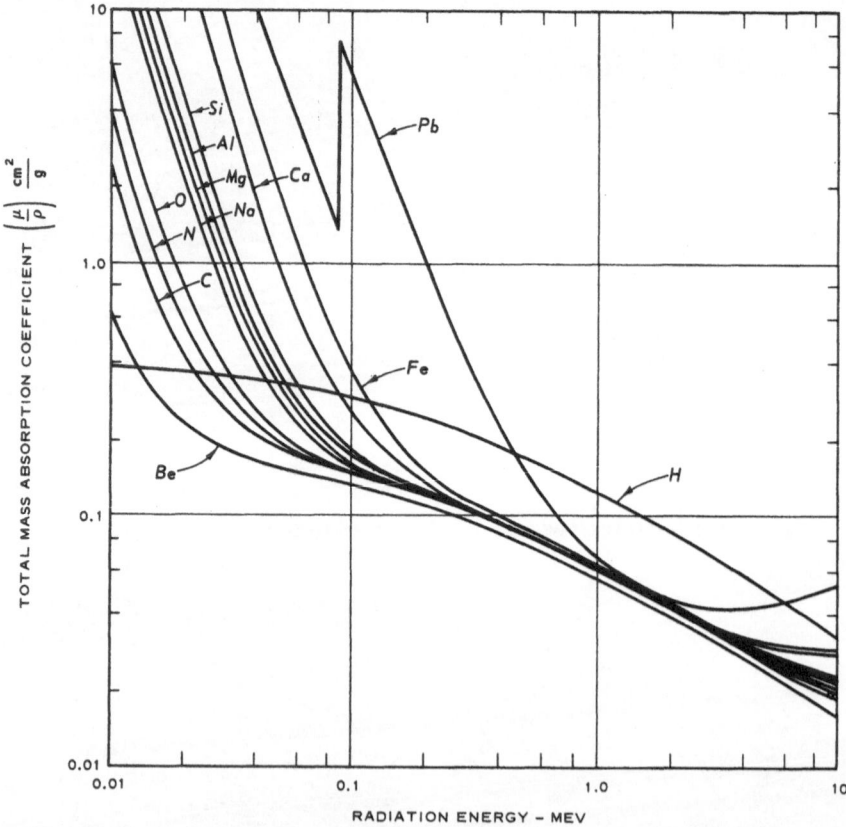

Fig. 11-5. Mass absorption coefficients *versus* radiation energy for selected elements (after Bloedow, 1962).

Table 11-1.   The Calculation of the Mass Absorption Coefficient μ/ρ for Kaolinite Irradiated by 300-keV X Rays

Element	Number of atoms	Atomic weight	Atomic weight in compound	R	Mass absorption coefficient of element at 300 keV	$R\left(\dfrac{\mu}{\rho}\right)$
Hydrogen	4	1	4	0.0202	0.212	0.0043
Oxygen	7	16	112	0.5657	0.107	0.0605
Aluminum	2	27	54	0.2727	0.104	0.0284
Silicon	1	28	28	0.1414	0.108	0.0153
Atomic weight of compound = 198				1.0000	$\left(\dfrac{\mu}{\rho}\right)_{compound}$ = 0.1085	

at 300 keV. Its mass absorption coefficient is developed in Table 11-1. With the determination of μ/ρ, the exponential decay equation can be solved to determine the change in intensity of the x-ray beam when it passes through a specimen of kaolinite.

For multiple layers, the equation may be stated as

$$I = I_0 e^{-\left(\frac{\mu}{\rho}\rho_T x\right)_1 - \left(\frac{\mu}{\rho}\rho_T x\right)_2 - \cdots - \left(\frac{\mu}{\rho}\rho_T x\right)_n}$$

where $1, 2, \ldots, n$ represent layers.

RADIATION PROBLEMS IN SOILS MODELS

The preceding sections suggest that the film density of radiographs may be measured by photometric means in order to determine soil density and also

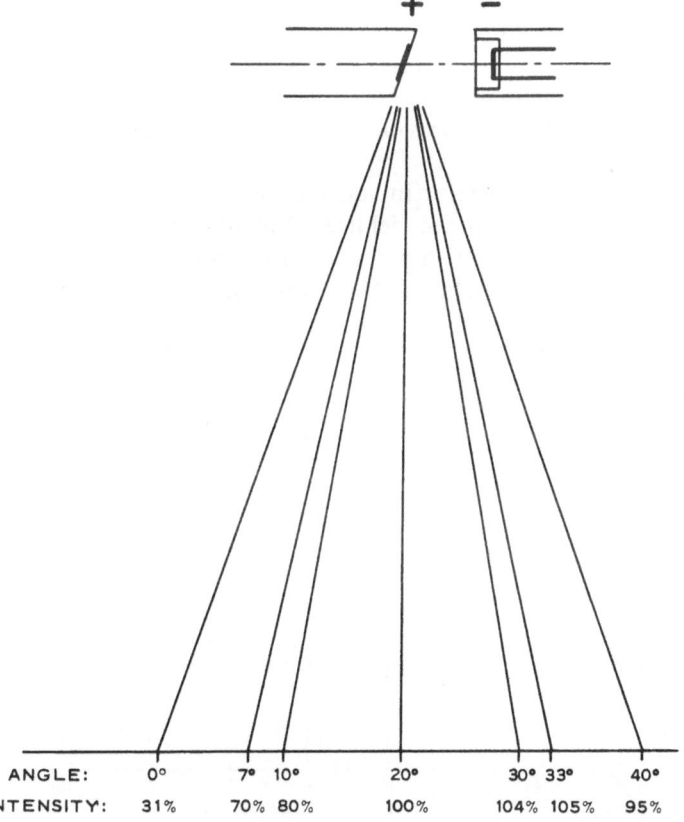

ANGLE:	0°	7°	10°	20°	30°	33°	40°
INTENSITY:	31%	70%	80%	100%	104%	105%	95%

Fig. 11-6. Variations of radiation intensity in terms of angle of emergence (after McMaster, 1959).

that measurements of radiation loss through a soil may, through calculation, provide measurements both of soil density and soil density changes induced, for example, through the process of soil testing. All this is true, but there are several sources of error which must be kept in mind.

Comparable radiation intensities may not be equitably distributed over the area exposed to radiation. Figure 11-6 illustrates what is called a "heel" effect in which a heightening of radiation intensity occurs in one portion of the area of radiation. In order to obtain comparable results, soil samples or models must be exposed in exactly the same position each time.

Image intensifiers should be used with caution where quantification of results is desired. Such images may have been intensified as much as 3000 times, during which process the image was reduced from about 7 in. in height to 1.5 in. and enlarged again to full size. In many models these changes produce a vignetting and dimension distortion, comparable to parallax effects in some cameras. Also, there is a decay in brightness along the margins of the viewing screen. Figure 11-7 shows a 40% decrease in image brightness. Thus, for quantitative work, direct exposure of radiation on film is best.

Flash x-ray equipment may fluctuate in radiation intensity from pulse to pulse so that density changes on film need not relate necessarily to soil densities.

A prime source of error in soil models is from radiation scatter and undercutting caused by the model itself. Figure 11-8 shows such effects in a uniform sand mass within an aluminum ring, contained, above and below, by square Plexiglas plates. Figure 11-9 shows a set of contours made from pho-

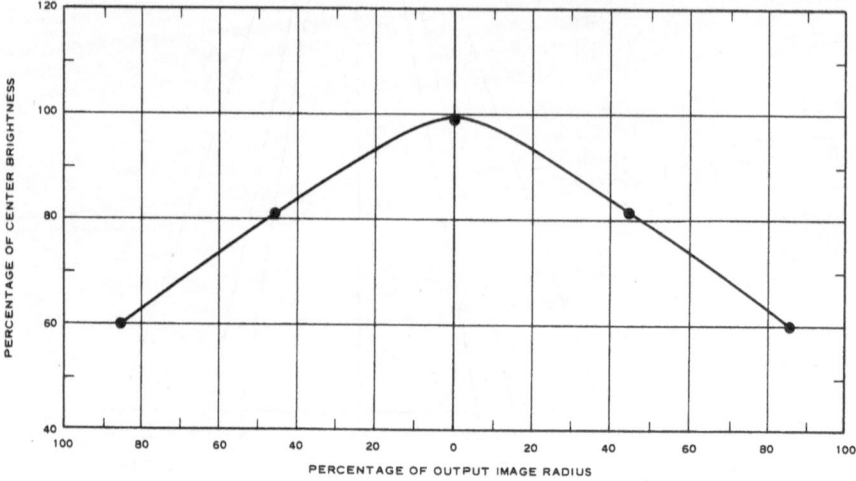

Fig. 11-7. Variation in image brightness across diameter of an image intensifier (after Baker and Janza, 1966).

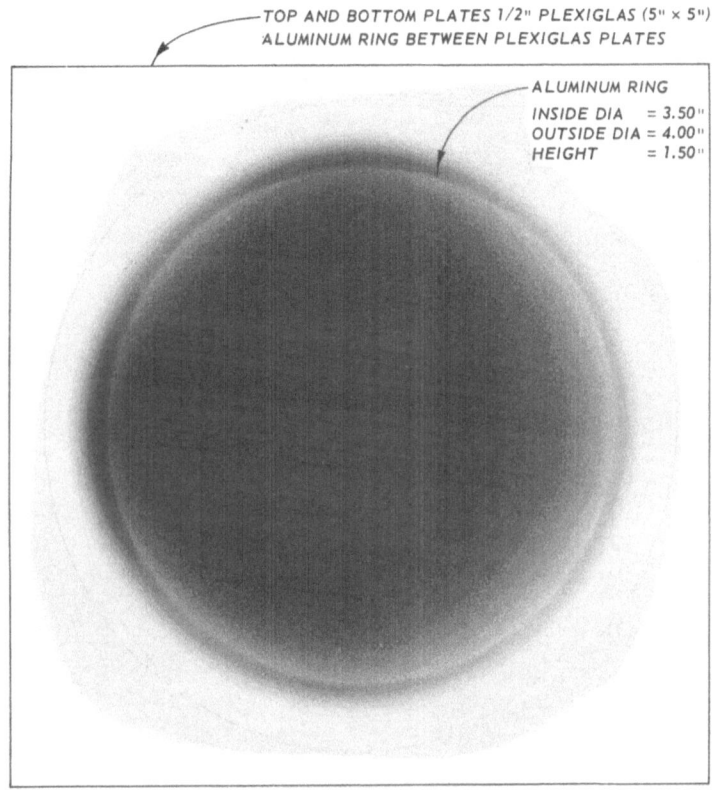

TOP AND BOTTOM PLATES 1/2" PLEXIGLAS (5" × 5")
ALUMINUM RING BETWEEN PLEXIGLAS PLATES

ALUMINUM RING
INSIDE DIA = 3.50"
OUTSIDE DIA = 4.00"
HEIGHT   = 1.50"

SAND DENSITY: 1.84 − FILM DENSITY: 0.51
9 MA, 69 KV, 60 SEC, 33-IN. F.D., KODAK TYPE M FILM
PATTERN OF RADIATION SCATTER
SAMPLE 13
OTTAWA SAND (20 TO 30 MESH)

Fig. 11-8. Effects of radiation scatter on a sand sample in an aluminum ring between square Plexiglas plates (direct print from radiograph).

tometric measurements. These measurements were point recordings made manually with a Macbeth photodensitometer. The values are in film density units. Note particularly that the contours have a square appearance, especially in the western half, reflecting increased radiation scatter introduced from the four corners of the Plexiglas plates. This radiation scatter undercuts the aluminum ring, which itself contributes a scatter effect greatly increasing the isodensity contours near the edge of the sample. The results show that care is needed in designing a soil test apparatus in order to minimize possible scatter effects which the apparatus itself may introduce.

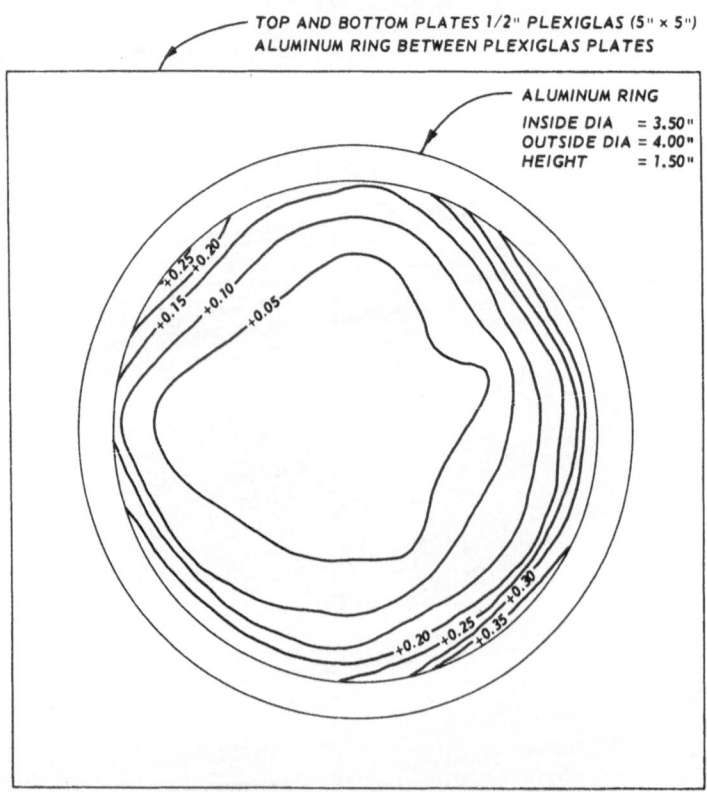

SAND DENSITY: 1.84 – FILM DENSITY: 0.51
CONTOURS INDICATE DEVIATION FROM BASE FILM DENSITY:
9 MA, 69 KV, 60 SEC, 33-IN. F.D., KODAK TYPE M FILM

PATTERN OF RADIATION SCATTER

SAMPLE 13
OTTAWA SAND (20 TO 30 MESH)

Fig. 11-9. Film density contours measuring deviation from base film density in Fig. 11-8. The contours may then be used as corrections.

Plastic sidewalls in a model, or sidewalls of materials with comparable low density, seem to cause the worst scatter effects in a specimen and the worst undercutting. Note the magnitude of such effects in Fig. 11-10. The contour values are three times those in the preceding example and the maximum deviation is over 0.90 of a film density unit. In practice, a correction of this magnitude is too great to make.

Aluminum or stainless steel rings are far better for sidewalls than plastic. Addition of a lead outer sheath is of additional usefulness, but the restraining plates through which the radiation passes need to be of Plexiglas or glass. Metal plates in the path of radiation cause scatter and too much loss of resolution.

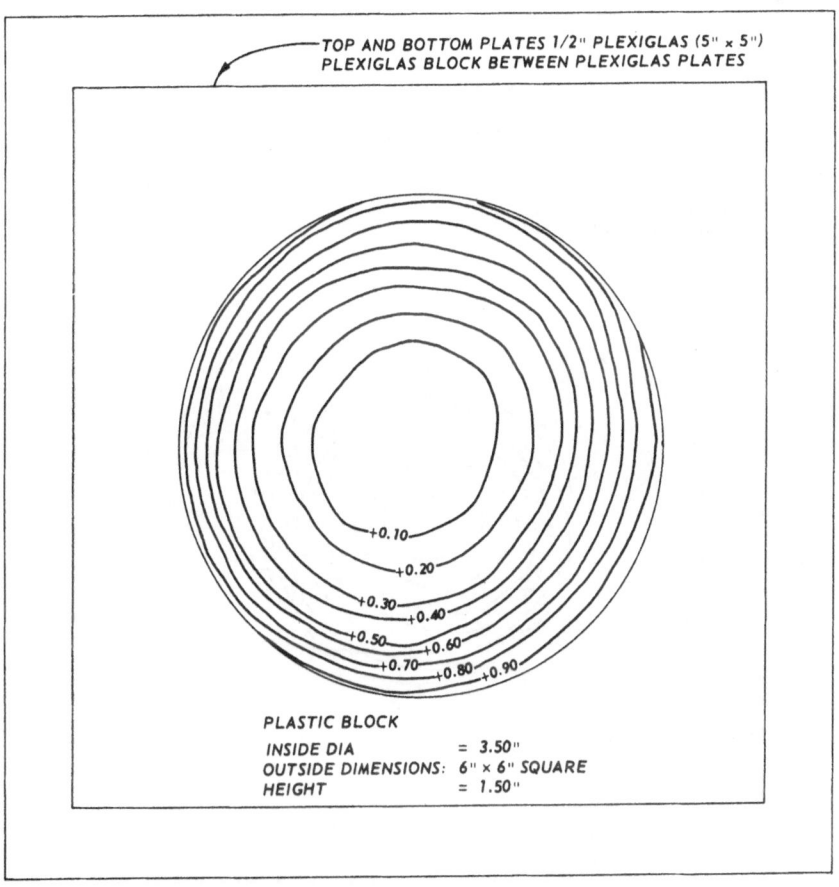

TOP AND BOTTOM PLATES 1/2" PLEXIGLAS (5" x 5")
PLEXIGLAS BLOCK BETWEEN PLEXIGLAS PLATES

+0.10
+0.20
+0.30 +0.40
+0.50 +0.60
+0.70 +0.80 +0.90

PLASTIC BLOCK

INSIDE DIA   = 3.50"
OUTSIDE DIMENSIONS: 6" x 6" SQUARE
HEIGHT   = 1.50"

SAND DENSITY: 1.74 – FILM DENSITY: 1.01
CONTOURS INDICATE DEVIATION FROM BASE FILM DENSITY:
9 MA, 69 KV, 60 SEC, 33-IN. F.D., KODAK TYPE M FILM

PATTERN OF RADIATION SCATTER

SAMPLE 87
OTTAWA SAND (20 TO 30 MESH)

Fig. 11-10. Effects of radiation scatter on a sand sample in a Plexiglas block between Plexiglas plates.

If feasible, the best way to reduce scatter in a model is by covering it on the radiation side with a lead sheet in which a window has been cut. The window needs to be smaller than the full area of the soil specimen so that the side-walls are not touched by radiation. In this case there is no scatter from the sides but there is a buildup of scatter from within the soil causing an increased radiation effect in the center of the sample and a relative decrease along the periphery. Figure 11-11 shows a pattern of this effect. Note the low order of film density deviation.

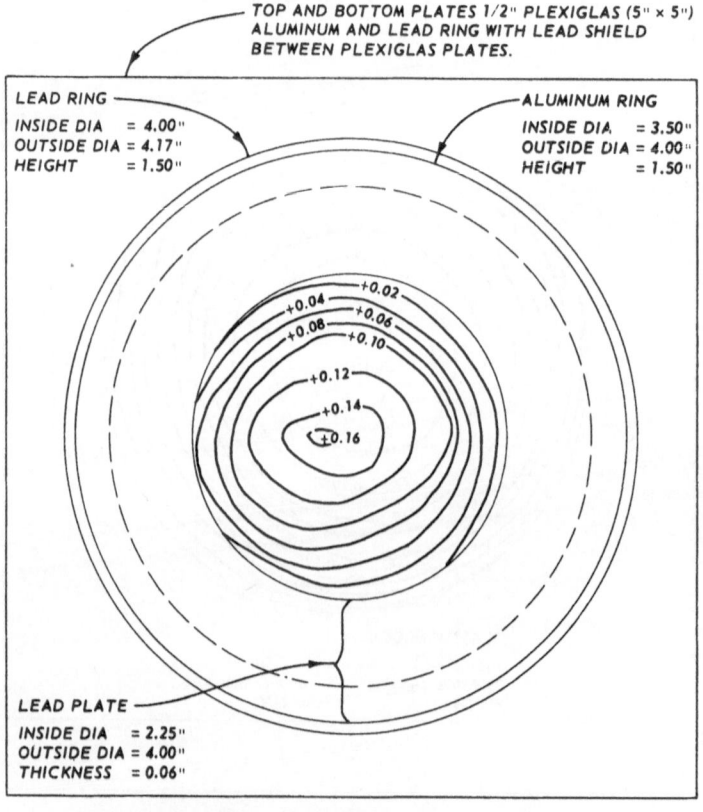

SAND DENSITY: 1.70 – FILM DENSITY: 0.98

CONTOURS INDICATE DEVIATION FROM BASE FILM DENSITY:
5 MA, 90 KV, 60 SEC, 33-IN. F.D., KODAK TYPE M FILM

PATTERN OF RADIATION SCATTER

SAMPLE 70

OTTAWA SAND (20 TO 30 MESH)

Fig. 11-11. Effects of radiation scatter on a sand sample radiated through a window in a lead sheet.

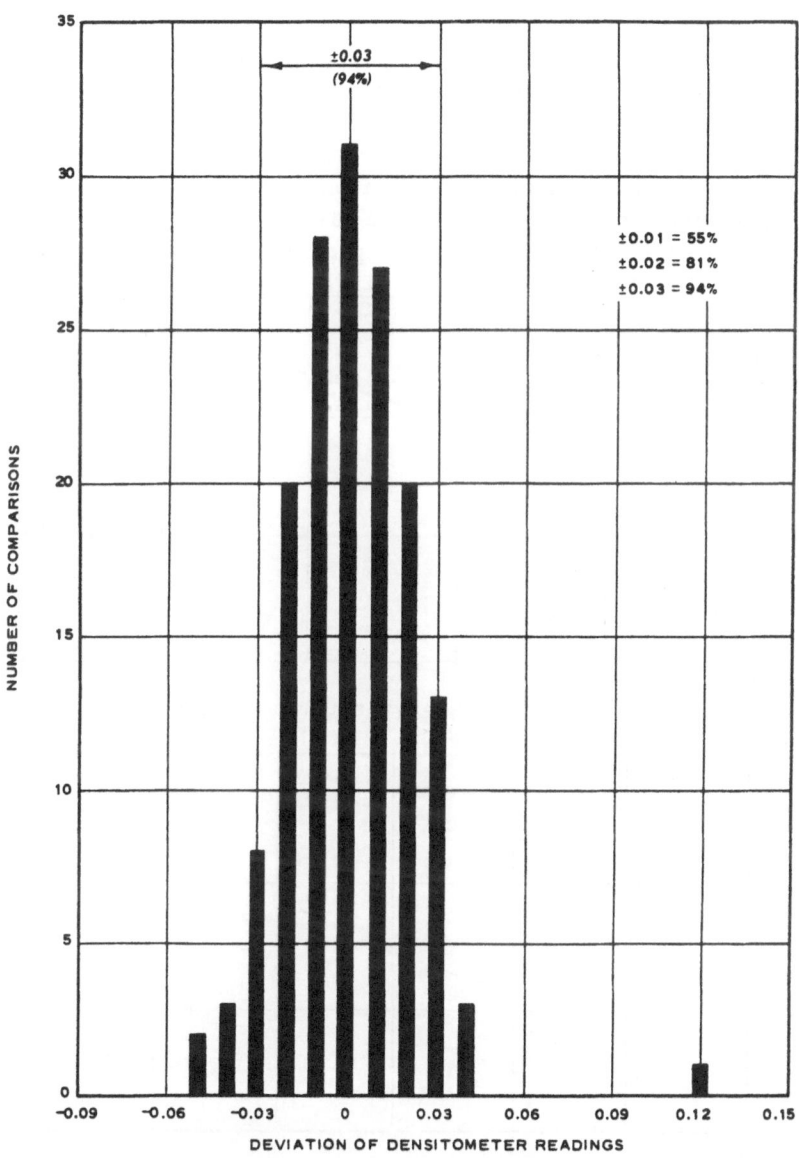

Fig. 11-12. Range of error found in a point-reading photodensitometer.

Thus, some of the possibilities for quantitative error in radiation experiments with soil models can be mitigated in the model design. The rest may be corrected for empirically. A useful corrective method is to prepare a control radiograph in which the scatter effects for a given model design and for the pattern of x-ray emission have been measured photometrically, the values then contoured to form a correction pattern for application to subsequent test radiographs.

The degree of chemical development of radiographic film directly affects the film density. It is good practice to run a standard film strip with each sheet of radiographic film used for quantitative analysis. These film strips should then be measured photometrically and compared with the value for the control

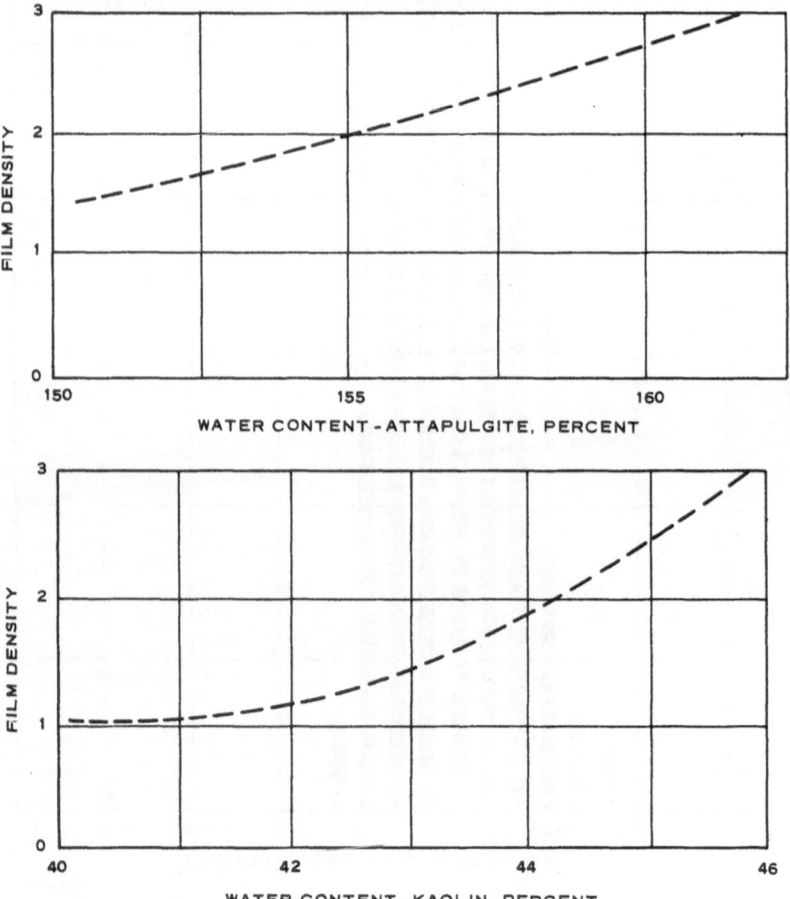

Fig. 11-13. X-ray film density *versus* water content in attapulgite and kaolin (after Leitch and Yong, 1967).

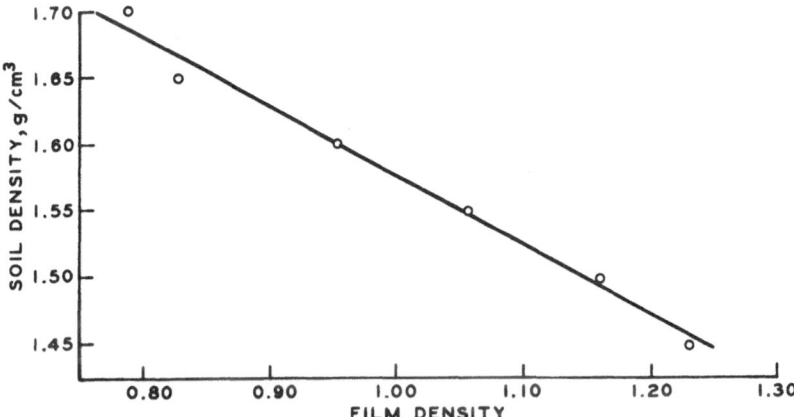

Fig. 11-14. X-ray film density *versus* soil density in Parafield loam determined by Greacen, Farrell, and Forrest (1967).

strip and a standard radiographic sheet. Measured density deviations may then be corrected for. However, if there is a deviation of 0.1 of a film density unit, or more, the radiographic exposure should be remade. Large corrections from film density strips tend to be unreliable but, in special cases, corrective relationships may be calculated (Ong, 1959).

Finally, the photometric device used for measuring film density may have its own range of error. Figure 11-12 shows the range of error in film density units by measuring and the remeasuring, a group of 156 standard film strips. Remeasuring was done on different days. The errors are introduced by degree of deterioration of the light source in the photodensitometer, changes in electrical current, though this is partly compensated for in most

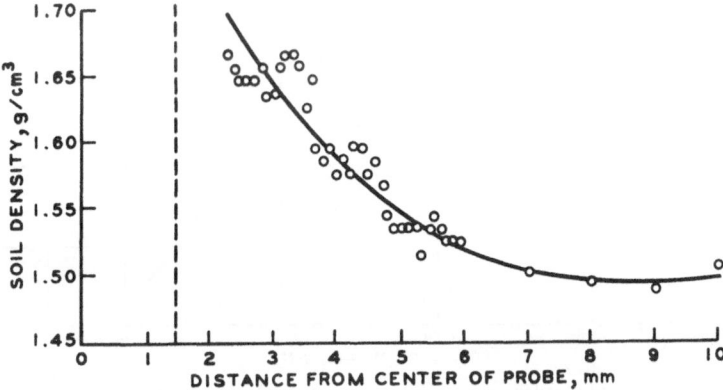

Fig. 11-15. Soil density determinations made with curve in Fig. 11-14.

units, condition of the apparatus, etc. Thus, error range cannot be attributed to specific models of photodensitometer units but should be determined periodically as work is being done.

SOIL DENSITY DETERMINATIONS

The preceding portions of this chapter suggest that image quantification should have many applications, particularly in soil studies. However, so far very little work has been done.

Leitch and Yong (1967) reported on film density relationships, produced by x radiation, as a measure of the water contents in attapulgite and kaolin (Fig. 11-13). However, their curves were prepared from calculations and they

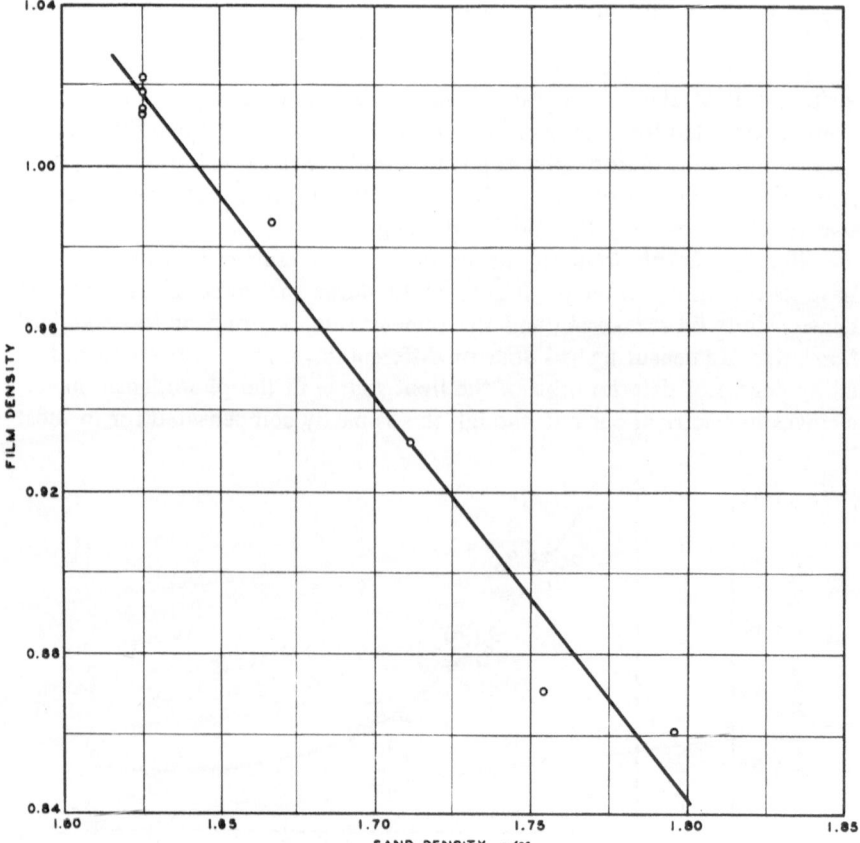

Fig. 11-16. X-ray film density *versus* soil density from Ottawa sand (from tests at Waterways Experiment Station).

varied the values for μ/ρ to accommodate varying proportions of H_2O in the clays. They also varied the densities of the clays to accommodate the changes in percentage of H_2O. Calculations were for sample thicknesses of ¾ in., radiation at 75 mA-sec and 25 kV with a 15-in. focal distance, using Ilford Ilfex x-ray film.

Radiographic measurement of induced density patterns in a soil (Parafield loam) has been reported by Greacen, Farrell, and Forrest (1967). As agricultural engineers they were concerned with the density changes in soils induced by the penetrations of roots. Their experiment was to push a metal penetrometer into a soil, impregnate the soil with a low viscosity epoxy resin, cut a slice and polish it to a thickness of 2.0 ± 0.01 mm, radiograph the specimen, and perform photodensitometer measurements on the x-ray film. Exposures were for 9 min at 15 kV and 20 mA, with a focal distance of 38 cm. The record was made on ordinary dental film. Other exposures on dental film were done with radiation from a cobalt source.

Figure 11-14 shows their film density versus soil density curve. The control curve was made from a stepped section specially prepared for this purpose. Figure 11-15 shows their interpretations of soil density changes in a radial path away from the probe. They report that this method is also being used for investigating density and crack patterns in ruptured soils.

Figure 11-16 shows an example of a calibration curve developed for an Ottawa sand (20 to 30 mesh) in experiments at the Waterways Experiment Station. Radiation was 69 kV and 9 mA for 60 sec, with a 33-in. focal distance, using Kodak type M film. The sample height was 1.50 in. and was contained between ½-in. Plexiglas plates. The curve was corrected for observable effects of radiation scatter. A group of these curves, for other sands, silts, and clays, would then serve for reference in model tests. Though such work has not yet been done, its practicability seems reasonably well established.

BIBLIOGRAPHY

Greacen, E. L., D. A. Farrell, and J. A. Forrest (1967). Measurement of density patterns in soil, *J. Ag. Eng. Res.* **12:** 311–313.

Leitch, H. C. and R. N. Yong (1967). The rate dependent mechanism of shear failure in clay soils, *Soil Mech. Ser. No. 21*, McGill Univ., Montreal, 140 pp.

McMaster, Robert C. (ed) (1959). *in:* Attenuation coefficient tables, *Nondestructive Testing Handbook*, Vol. 1, part 27, The Ronald Press, New York, pp. 1–41.

Ong, Sing Poen (1959). *Microprojection with X rays*, Martinus Nijhoff, The Hague, 132 pp.

INDEX